Jai McKenzie is an artist and academic, originally from Sydney, she now calls Berlin home. She has lectured in Electronic and New Media Art at Sydney College of the Arts, photography at the University of Technology, Sydney and the University of Newcastle, Austalia.

LIGHT+ PHOTOMEDIA

a new history and future of the photographic image

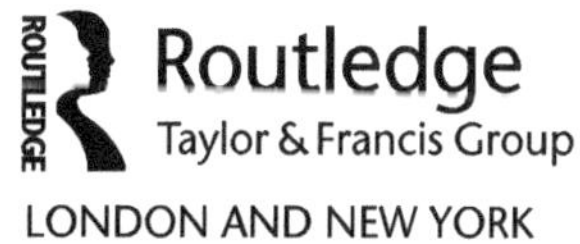

LONDON AND NEW YORK

First published 2014 by I.B.Tauris

Published 2020 by Routledge
2 Park Square, Milton Park, Abingdon, Oxon OX14 4RN
605 Third Avenue, New York, NY 10017

Routledge is an imprint of the Taylor & Francis Group, an informa business

The International Library of Visual Culture 8

A full CIP record for this book is available from the British Library
A full CIP record is available from the Library of Congress

Library of Congress catalog card: available

Printed and bound in Great Britain by TJ International Ltd, Padstow, Cornwall

ISBN 13: 978-1-7807-6278-4 (pbk)

Contents

List of Illustrations

INTRODUCTION

In this book, I propose that the history and future of photomedia are fundamentally connected to light. I argue that, regardless of technological change, light is a constant defining characteristic of photomedia intrinsically coupled with space and time to form explicit light-based structures and experiences. At times throughout the book, the three primary elements – light, space and time – are dealt with as individual entities in order to understand the mutual and evolving connections between them.[1]

The word photomedia is often used within the confines of media and art theory, as an abbreviation of the two words photographic and media. The term refers to the scope of photographic and video technologies that are used and critiqued within those fields. In this book the photographic devices that fall under the term photomedia are as broad as the etymological scope of the words photographic and media. Essentially I consider all devices that use light and media as photomedia including photography, cinema, video, television, mobile phones, computers and photocopiers. Thus for reasons of brevity I have not dealt with every photomedia device and primarily discuss early photographic devices such as the camera obscura, analogue photography, film, video, and digital photography and film. The focus of this book has been largely kept within the canon of photomedia, with a strong emphasis on scientific, fine art and conceptual and contemporary art practice. But again while this discussion does not discount the broader field of image production it does not, for reasons of brevity and clarity, discuss every aspect.

The book structure reflects the four technological eras: *Early image machines* c.1830 to c.1870, *Analogue image machines* c.1870 to c.1990, *Digital image machines* c.1990 to the present day and *Future image machines* of the year 2039: approximately two hundred years after the invention of photography. The light-space-time structures are analysed as they continuously develop through each of these characteristically different epochs.

Because of the long timeline which frames this book, no single artist or case study is consistently used; rather, representative artists from each era are discussed in terms of the light-space-time structures they construct.

While the specificities of the technological history and futurology of photomedia are not central to this book, they are relevant to the context and support the arguments concerning the technologising of light. Regarding the technological aspects of this book, no single source for the complete history of photomedia was found. Nonetheless, some excellent references provided useful background for the compilation of the techno-historical aspects.[2] Information on the most contemporary developments from 1990 to the present were researched through specific photomedia manufacturers' websites and publications,[3] relevant essays and enthusiasts' blogs and databases, however many of these sources were biased or inadequate and required a large amount of cross-referencing. I also found that despite its almost 20-year presence and probable future importance the body of literature on digital photography is limited. The speculative considerations of future technological change are based on analysis of an enduring and widely applied technological growth theory known as Moore's Law.[4]

The scant amount of material dealing directly with photomedia and light encouraged me to discuss each of these two topics in a new way. In an effort to mark out some territory in this field I have drawn upon the theories of Jean Baudrillard,[5] Vilém Flusser[6] and Paul Virilio.[7] In addition, I have also purposely dealt with the progression of light-space-time structures over the full extent of the historical timeline to identify the progression of these structures throughout photomedia's history and future.

Chapter 1, 'Early image machines: the invention of photography, c.1830 to c.1870', investigates the emergence of photomedia with a focus on the development of image machines and their direct impact on the structure and experiences of light-space-time. This most luminous era is analysed through discussion of the period that provides the historical reference point for the invention of photography, as well as the 40-year period after the official recognition of photography in 1839. The first part of the chapter briefly outlines the social conditions necessary for the invention of photography. Here I discuss the camera obscura and various chemical light storage systems. Next, an analysis of the impetus for the invention is given through a discussion of the scientists (artists) who, through their technological enquiries into light, produced the earliest photographic images. The scientists who invented photography were also the artists who explored the new visual possibilities the technology could offer with the remarkable emergence of these early image machines. William Henry Fox Talbot, Louis Jacques Mandé Daguerre and Joseph Nicéphore Niépce are generally acknowledged to have made major contributions to realising the invention; they developed the technology and used it to produce some of the earliest photographs. Some of these images are analysed in terms of their associated impact on experiences of light-space-time. I discuss Daguerre's *Boulevard du Temple, Paris,* c.1838, Talbot's *Latticed Window,* 1835, and Niépce's *View from the Window at Le Gras,* c.1826 which demonstrate a new photographic time shift where light-time is compressed in the stillness of the image. Virilio's theory of 'dromoscopy'[8] is discussed in relation to speed and movement, light and photography. Baudrillard's theory of the simulacra and Flusser's theory of the apparatus and the ensuing 'photographic universe'[9] are introduced in relation to the earliest examples of photomedia. Many other image makers contributed to the body of work I call light-space-time constructs. Notable examples are Hippolyte Bayard and his image *Self Portrait as a Drowned Man,* 1840, John Whipple with *The Moon,* c.1840s and the later image *Four diameter cross-section of segments of the cerebellum* by Dr Jules Bernard Luys, c.1870. Although Bayard, Luys and Whipple are not credited with the invention of, or major developmental contribution to the early era of photomedia, their works provide telling insights into the early light-space-time structures. These works exemplify the new image-spaces which formed through light-space, where distant objects are brought close and internal understanding is manifest externally through the image. Through the works of all these early photomedia practitioners, I argue we see the development of what I term image-space. That is a space of images which exists in the shared external, physical world as well as the inner psychological world of each individual who engages with photomedia. Overall, this chapter provides a new rationale for the invention of photography in which its fundamental relationship to light leads to the redefinition of photographic images as light-space-time structures enabling the emergence of the 'image-space' constructed through the reproduction and reception of photo-images as they begin to proliferate through society as image-worlds.

Chapter 2, 'Analogue image machines, c.1871–c.1990' covers a period spanning more than a hundred years and one in which technology evolved faster than any century before it. Rapid broad-scale technological change characterised the analogue era, which began with locomotive transport and ended with jet aircraft and space shuttle launches. The light-space-time formations which glimmered in the earliest era shone much brighter in this period in which photomedia became a tool for exploring and expressing a broad range of experiences and concerns spanning science, sociology, ontology, phenomenology and epistemology. While this era is enormously rich with material for discussion, for reasons of brevity and conciseness, I have focused my research on a selective cross-section of artists using photography, cinema and video who clearly demonstrate the progressive changes in light-space-time and the increasing proliferation of image-spaces which began to dominate mass media. Although it is discussed, I do not dwell on television within this era as it has been well researched. Certainly, discussions concerning the theories of Baudrillard and Virilio reference televised imagery.

I begin the analysis of analogue light-time structures through a discussion of chronophotography[10] including the work of Eadweard Muybridge and his image *The Horse in Motion,* 1878, Étienne-Jules Marey's *Movements in pole vaulting,* c.1900 and Harold Edgerton's *Bullet through Banana,* 1964. Speed entered light-time structures leading to an instantaneity exemplified and discussed through the aforementioned works. The deconstruction of physical movements through space and time was captured through photography for later visual reception. I then move to the antithesis of chronophotography – Photofuturism and Anton Giulio Bragaglia's *The Typist,* 1911, a work which demonstrates a free flowing and chaotic expression of physical movement through space and time. Following this is a discussion of light-space-time specific to cinema based on Lászlό Moholy-Nagy's *Lichtspiel Schwarz-Weiss-Grau,* 1930, a work which exemplifies the supremacy of light as Maholy-Nagy himself espoused. The works of Michael Snow's *Wavelength,* 1967, Chris Marker's *La Jetée,* 1962 and Andy Warhol's *Empire,* 1964 provide the basis for a discussion of the parallels between the still and moving image in mid-twentieth-century structuralist cinema. David Campany's concept of cinema initiating stillness[11] is also considered and the application of Virilio's dromoscopic model to speed and light within cinema is further explored.[12] This is followed by a discussion on the development of light-space and image-space in cinematically influenced photography such as Cindy Sherman's *Untitled Film Still,* 1978 and Hiroshi Sugimoto's *Ohio Theatre, Ohio,* 1980. Further examples of image-space and their relationship to Flusser's theory of photographic universes[13] are considered through, for example, the documentary photography of Henri Cartier-Bresson's *Place de l'Europe, Paris,* 1932. The structure of fragmented image-space and the methods employed by artists to express this, in works such as John Heartfield's photomontage *Hurrah, die Butter ist alle!,* 1935 and Martha Rosler's *Red Stripe Kitchen,* 1967–72 is explored, as is the resistance against the photographic apparatus as theorised by Flusser.

Early video art and related light-space-time is also considered in relation to the image-space in an analysis of Nam June Paik's *Moon is the Oldest TV,* 1965, Andy Warhol's *Outer and inner space,* 1966 and Buky Schwartz's *Yellow Triangle,* 1979.

Chapter 3, 'Digital image machines', investigates the technological era from 1990 to the present. The focus of this chapter is the light-space-time structures which emerged in the 1990s with the introduction of digital photomedia which forms the primary method for photomedia today. The chapter starts with an analysis of theories of the void in relation to digital technology and the perceived loss of physicality in this most recent epoch. Next, a brief discussion of the short technological history of digital photomedia is provided, followed by an exploration of the light-space structures that the use of this technology has initiated. Examples used are Jeff Wall's *After 'Invisible Man' by Ralph Ellison, the Preface,* 1999–2000, Gregory Crewdson's *Untitled 'Beneath the Roses',* 2004 and Charlie White's *Champion,* 2005. As part of this enquiry into the light-space structures of the era, screen spaces are discussed. Then, the dynamic speed of light-time and speed's antithetical expression, slowness, are analysed in terms of Virilio's theoretical frameworks of 'dromology' and 'dromoscopy'. Denis Beaubois' *The fall from Raiatea,* 2007, Gillian Wearing's *Snapshot,* 2005 and Bill Viola's *The fall into paradise,* 2005 exemplify the paradoxical accelerated yet relative slow relationship of time. Finally, a consideration is made of the manifestation of light-space-time in David Noonan's *Kabarett Keif,* 2007, Andro Wekua's *Lying, Walking, Swimming,* 2005 and Daniel Crooks' *Train No. 1,* 2002–5. These works demonstrate the fractured nature of image-space and are discussed in relation to Baudrillard's understanding of contemporary communication and fragmentation,[14] Virilio's concepts of fragmentation and geometry,[15] and Flusser's theories of the 'photographic universe' and the function of the apparatus.[16]

Chapter 4, 'Future image machines 2039', speculates on the as yet unrealised potential of photomedia. The central points of enquiry are the effect of two hundred years of photomedia in the year 2039 and the related light-space-time structures. The chapter begins with a consideration of the possibility of predicting the future and the utopian and dystopian perspectives that influence such speculations. An overview of Gordon E. Moore's model for calculating technological growth is given as the basis for predictions of the increasing ubiquity of photomedia and how artists will use and respond to this environment. The effect of this environment which is saturated with photomedia images is discussed in terms of the 'photographic universe' and the apparatus as discussed by Flusser,[17] Baudrillard's concept of the void[18] and Virilio's analysis of speed and light.[19]

Overall, this book provides a history and futurology of photomedia and its fundamental relationship to light. The approach aims to establish a new, expanded perspective on the medium through the introduction and analysis of the concepts of light-based space and time structures. Such an analysis leads to a greater understanding of photomedia and its significant impact on contemporary and future experience and perception.

1

EARLY IMAGE MACHINES: THE INVENTION OF PHOTOGRAPHY C.1830–C.1870

Luminous beginnings

Was the era from c.1830 to c.1838 'completely bathed in luminous fluid'[1] as the inventor of the first semi-permanent images, Joseph Nicéphore Niépce[2] contended? Possibly. Within the radiant beginnings of early photography, an understanding of light was central to the development of the relevant photochemical and mechanical inventions required to realise it. During this epoch, in which the invention of photomedia emerged, light attained a palpable presence within society.

In Western philosophy and science there have been varying historical interpretations of light. However, one constant idea persists: the supremacy of light within the totality of experience. The nature of light is understood in terms of its physical properties, and its theoretical relation to experience. We engage with light in many ways, visually and physically, emotionally and intellectually. Light underpins visual perception and therefore is the critical source responsible for our understanding and orientation in space and time.[3]

Visual experience is impossible without light; our eyes receive it, allowing the passage of light to travel from outside to inside the body. In simple terms, light enters the eye as photonic information, it strikes and is subsequently collected by the retina at the rear of our eyes. Here the eye functions as a converter, transforming light into a complex series of electrochemical events in the retinal receptors generating electrical impulses that are transmitted down the optic nerve to the brain, which aggregates and interprets the electrochemical information it receives, creating neural constructs that form images in the final process of visual cognition. We see and understand the world through this fundamental relationship between light and visual perception.[4] Importantly, because of the finite speed of light and the delay that occurs as light is transformed into electrical impulses that move through the brain, we always sense the recent past.[5] Likewise, photomedia secures light from the past as visual imagery, bringing it to the present for further reception. Both methods of recall occur due to either psychological or physical light characteristics. Through the cognitive processing of light we seek to understand the universe. Through light, photomedia was formed.

Light source: the origin of the image machines

Any discussion on the development of photomedia technology necessarily begins with the camera obscura.[6] The principles of the camera obscura were first observed by Aristotle during an eclipse of the sun.[7] This proto camera device, which functioned by using a dark chamber fitted with an aperture, captured light but did not have the capacity to store an image. It required an artist to render the scene by hand. This was a significant limitation that demanded a solution. It is widely acknowledged by historians that for the invention of photography to occur there needed to be an effective control of light through an instrument like the camera obscura working in combination with an appropriate storage process.[8] As early as 1727 the German anatomy professor Johann Heinrich Schulze[9] observed that silver nitrate darkened when exposed to light.[10] Although Schulze recognised that chemicals could provide an appropriate storage device for light he considered his observations relevant only in testing for the presence of silver in alloy metal.[11] Nevertheless, these observations emphasise the importance of light within the discovery of photography and ultimately form the methodological basis of black and white photography.[12]

Despite this knowledge of both the camera obscura, an instrument for controlling light, and silver salts, the essential chemical processes to 'store' light, it was not until the late eighteenth century, at the beginning of what is known as the Industrial Revolution, that the fertile ground for the growth of the image machines came about. Consistent with the growth of industry, and because photography was seen as a mechanical process, in France people spoke of the first general process of photography, the daguerreotype, as *la machine Daguerre.*[13] Meanwhile in England, scientist and photographer Sir John Herschel[14] coined the term photography, which literally means 'light writing'.[15]

The first recorded experiments that aimed at producing photographic images were made by Thomas Wedgwood[16] between approximately 1790 and 1802 using a camera obscura and nitrate of silver to produce images. Although Wedgwood made great advancements towards developing a process of photography, he failed to achieve a method of fixing the images once they were exposed. It was not until 30 years later that William Henry Fox Talbot[17] made progress on this aspect of the invention. One of the most well known inventors of photography, due to his substantial output of photographic images,[18] Talbot speculated on the idea of photography while using a camera obscura, remarking, 'how charming it would be if it were possible to cause these natural images to imprint themselves durably, and remain fixed upon the paper!'[19] Talbot's understanding of the light-based camera obscura images informed his awareness of the antithetical property of light, the shadow. By 1835, Talbot had developed a photographic system of depicting objects through their shadows, which he called skiagraphy.[20] With this working negative system Talbot understood the potential application of the negative object to make a positive print. However, he had not devised a way to fix these images to paper until a month after Louis-Jacques-Mandé Daguerre's[21] invention was announced.

But, as Herr Schiendle suggests, we might consider the first inventor of photography to be Johann Bapista Porta who popularised the camera obscura, or as Schiendle states, 'with equal reason to whoever first noticed the fading of a curtain under sunlight'.[22] For on this line of reasoning, 'the discoverer of a fact is also the discoverer of whatever knowledge that fact may ultimately lead to.'[23]

In any event, as Peter Galassi points out 'the most curious aspect of the race to invent photography is that it was not a race until it was over.'[24] Galassi claims there was no clear inventor of photography and that, depending on one's definition of photography and the characteristics of the medium one finds paramount, the inventor could be Wedgwood in 1802, Niépce in 1826, Talbot in 1835 or even Daguerre in 1835 or 1839.[25]

Regardless of which one of these educated and financially secure visionaries of the early Industrial Revolution invented photography, they all manifested the sociological, artistic and scientific aim of the era: to accurately transcribe the physical world. In essence, each sought a means of writing with light directly from nature.[26]

A sociological perspective

For the most part, historical accounts of early photography fail to deal in depth with the multifaceted aspects of photography's history. In general, these accounts discuss the technical conditions or closely examine the individuals associated with the invention. They do not, however, consider the deeper philosophical aspects of the period let alone the significance of light. However, most acknowledge the importance of the historical period between 1830 and 1839 in the invention of photography. As Peter Galassi states, 'even the driest technical histories implicitly acknowledge that photography was a product of shared traditions and aspirations.'[27]

The most thorough contemporary exploration of the sociological aspects which contributed to the impetus for the invention of photography is *Burning with Desire: The Conception of Photography* by art historian Geoffrey Batchen. He argues that photography was brought about by particular cultural and social contexts within European history.[28] He states that

> photography appears to have emerged as the embodiment of certain Western knowledge rather than as a creative "idea" or technological "discovery" traceable to the actions of any particular individual. Photography, it seems, was a product of (and contributor to) certain shifts and changes within the fabric of European culture as a whole.[29]

In addition, Herbert Ohlman claims that any invention cannot succeed without certain prerequisites; there must be entrepreneurs willing to fund development, and there must be an unmet need for the invention.[30]

Jean-Claude Lemagny and André Rouillé suggest in their book *A History of Photography: Social and Cultural Perspectives*[31] that the development of photography was, since the Renaissance, driven by a quest for accurate resemblance within Western art; a search for the perfect facsimile that strove to both arrest a gesture and suspend time.[32] They postulate that this point in the history of art coincided with a rising middle class who wanted to be able to control everything possible in an effort to discover the potential marketability of all things. The invention of a device for 'cataloguing the world', so to speak, was a valuable asset because 'without this there could be no accurate inventory or any complete investigation of the world which this class was now bent on controlling.'[33] In addition, Peter Galassi points out that the great political and social transformations of the industrialised period fostered many speculative thinkers, whose ideas and output provided answers to such requirements but subsequently introduced further needs.[34] For instance, the Wedgwoods benefited from the industrialisation of England and capitalised on it through automating mass produced pottery. As Tim Cloudsley remarks, an 'increase in population in eighteenth century Britain and a slight rise in the average standard of living, supplied a minimal domestic market for the first mass-produced goods.'[35] It is highly probable that Wedgwood's motivation for experimenting with photography was in part an attempt to assist in his father's mass production of pottery, by replacing hand painted surface imagery with photographic imagery. Wedgwood would have understood the potential commercial applications of photography and clearly perceived the possibility of a direct application to pottery.

However, the accelerated pace of industrialised society created anxiety; an apprehensive environment in which an opposing desire for stillness developed. Without doubt there was a propensity to try and understand the rapidly changing world through the stillness of an image. Photography satisfied that desire, as it allowed viewers the opportunity to linger over a static image of a moving world. The image was considered to be truth. A faithful, useful and comforting representation of a moment of the real was seemingly found in photography. Laura Mulvey locates this satisfaction within an 'aspiration to preserve the fleeting instability of reality and the passing of time in a fixed image'.[36] But, Mulvey, like Batchen, Thomas and others, fails to acknowledge the importance of the scientific and poetic exploration of light in the invention of photography. As Ann Thomas affirms, 'a creation of both science and art, photography owes its physical existence to the research of the chemists, opticians and philosophers of light of the eighteenth and nineteenth centuries.'[37]

Light-space-time

Wedgwood's interests in the application of photographic techniques did not arise exclusively from commercial pursuits but also from his research into light, visual perception, education and learning. Wedgwood, when discussing his studies of educational techniques, once said his aim was 'to find some master-stroke which should anticipate a century or two on the lazy-paced progress of human improvement'.[38] Through close study of infants, he observed that most of the information a child received related to light taken in through the eyes. This realisation led him to consider the possibilities of images created by light. Wedgewood's understanding of the relationship of light and imagery to comprehension and perception was an astute observation. Clearly, Wedgwood was conscious of the power of light-based images to inform our understanding of what we perceive. He saw the potency images had as carriers of visual information and consequently that images could be articles of knowledge. His subsequent quest to bring about a photographic process was part of a larger desire to understand the relationship of light, imagery and knowledge; it was not simply a commercial venture.[39]

While Wedgewood (and others) can arguably lay claim to inventing photography, 7 January 1839 is the indisputable point that the process devised by Daguerre and Niépce was announced by François Arago to the Academie des Sciences in Paris. Within a matter of days, the announcement was officially published for all to know.[40] Earlier, Niépce, working alone, had attempted to have his invention recognised by the Royal Society but was rejected, possibly because of a lack of detail in his report concerning the process.[41] After this disappointment, Niépce joined with Daguerre to perfect the process but died suddenly in 1833. Although Niépce had brought a large portion of the scientific knowledge of photography to their partnership, Daguerre finalised the process and convinced the Academy of its unique value. This left Daguerre and the widow of Nicéphore Niépce as the recipients of a financial subsidy supplied by the French government for publishing the photographic system that Daguerre had self-referentially titled the Daguerreotype.[42] From the moment the process was published, it was publicly available and was immediately taken up by a large portion of the bourgeoisie and artists.[43] The popularity of the device was such that 'Daguerreomania' was reported by the French press. The fervour moved quickly from France to the United States but did not proliferate through England at the same pace.[44] This was largely due to the English entrepreneur Richard Beard,[45] who bought the patent for the country and tightly controlled copyright; however, this did not hold it back for long.

In 1839, the same year that the Daguerreotype was announced, Daguerre proclaimed, 'I have seized the light.'[46] His aim was to capture 'the spontaneous action of light'.[47] At the time of his image, *Boulevard du Temple, Paris* (fig. 1.1) Daguerre was exposing

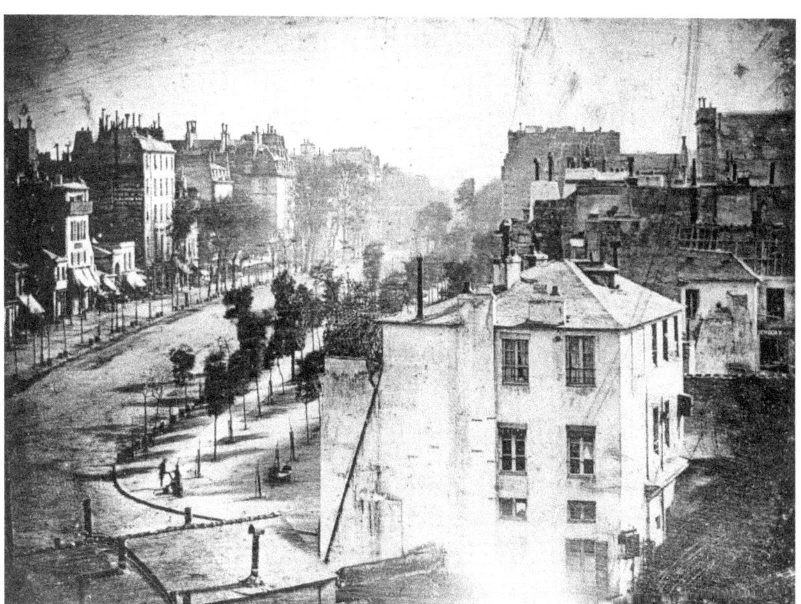

Figure 1.1 Louis Jacques Mandé Daguerre, *Boulevard du Temple, Paris*, c.1838.

light sensitive material for approximately 60 seconds, rendering invisible any movement that would normally have been seen during the duration of the exposure. The view Daguerre chose to capture would have certainly originated from his apprenticeship in panoramic painting.[48] This is evident in Daguerre's compositional approach; he framed the image as any painter would, for a picture perfect view of the Boulevard. But, *Boulevard du Temple* does not depict the bustling Parisian boulevard that would have been unfolding before Daguerre's lens during the exposure, unlike Claude Monet's 1873 painting *Boulevard des Capucines* which Monet painted, interestingly, from the window of Nadar's photographic studio.[49] Daguerre's photograph only depicts the man who coincidentally stops to have his shoes shined during Daguerre's exposure. This image is of profound historical importance: it is the first known photograph to record a human presence.[50] Due to the man's lack of movement over time, his presence materialises in the bottom left side of the photograph. Not only does this image document a moment in history, but that moment which Daguerre chose to photograph also extends our understanding of light-time; that moment becomes an elongated and present light-space. Through the depiction of the *Boulevard du Temple,* Daguerre forms an image entirely by the capture of light. *Boulevard du Temple* demonstrates the ability of the photographic image to represent phenomenological time; it is an indicative snapshot of the era, as that representation of time is not only slow, it is still. Carried forward with the image itself, time is contained and sustained in light-space.

Boulevard du Temple captures a photographic time shift, where the experiential space of the image produces a compressed light-space-time. When viewing this image we are shifted from the usual experience of time and space; the 'now' of the photograph is simultaneously an elongated, durational, past time.

Daguerre may have 'seized the light' however, arguably, the most poetic inventor of photography was Niépce. He had a unique understanding of light which he applied to the invention of his photographic process. In a note dated 5 December 1829, Niépce informs Daguerre that, 'in the process of composing and decomposing, light acts chemically on bodies. It is absorbed, it combines with them and communicates new properties to them... This, in a nutshell, is the principle of the discovery.'[51] Niépce's claim suggests to me that he understood light better than any other at this time, as he understood both the possibility to convey information through objects with light and its transformative abilities.[52]

View from the Window at Le Gras (fig. 1.2) provides us with insight into Niépce's intimate understanding of the qualities of light. To me it is clear that *View*... precisely demonstrates that, when harnessed in service of photomedia, light conveys new space and time perspectives. *View*... was shot circa 1826 from a window in Le Gras in Saint-Loup-de-Varennes.[53] The image displays a metallic luminescence, no doubt due to the specific materials used in making the image. Niépce's photographic system, which he named heliography,[54] involved exposing to light a substrate of tinplate coated in bitumen. His use of materials and technique resulted in a metallic purple-black image. All of Niépce's early heliographs were made over a protracted period of time; *View*... required an eight-hour exposure. The result is a photograph that depicts a 'view' of time and space that observers, not recognising the value of the temporal dimension, might think of as distorted.

Niépce's day-long open shutter creates a lasting, tranquil and illuminating visual experience. The long light exposure impresses a luminescent permanence on the sensitised substrate; this day is a day that seemingly never ends. Through those eight hours of light and shade, most of this day is rendered visible. Specifically, shadows fall on both sides of the gabled roofs due to the long exposure, creating a peculiar effect that is not normally seen. Movement however is not visible; even the sun, which imperceptibly moves from the eastern horizon to the western horizon, travels too quickly for its image to be captured. At the moment Niépce developed this photograph he may well have recognised that this image represented these features of light based expressions of space and time.

The temporal nature of this photograph is not restricted to that one day in 1827. Like all photographs, Niépce's *View*... has travelled and continues to travel through time and space, bringing past time and distant space into the present. This is not light as we normally experience it. As Paul Virilio explains, 'the photographic

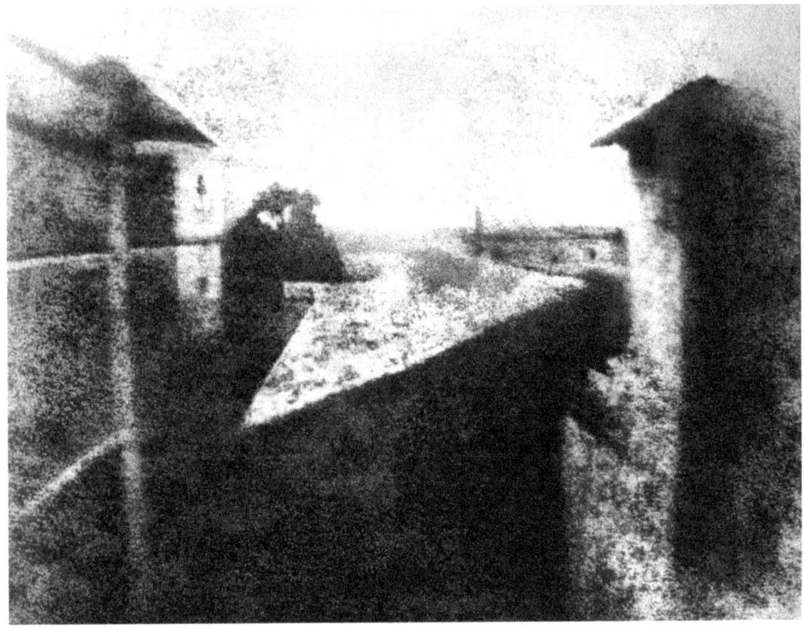

Figure 1.2 Joseph Nicéphore Niépce, *View from the Window at Le Gras*, c.1826.

exposure lends daylight a temporal measure independent of the meteorological day.'[55] In other words, the photographic image functions in a temporally specific manner, this is a new time, a photographic time. *View...*, like *Boulevard du Temple,* physically demonstrates that photography, from its earliest beginnings, binds light with time and space, communicating new properties of each. At this fundamental stage, the novel characteristics of these light-space and light-time structures are that light-space transposes and transports space while light-time slowly impresses itself as a two-dimensional image (which also travels through time).

Paul Virilio remarks on the luminance of Niépce's early images, and the way light gives depth and weight to the image. He suggests that the remarkable quality of the heliograph is that these 'solar writings' allow light and shade to be 'impressed' on them.[56] This is the beginning of what Virilio terms the 'industrialisation of vision' which occurs when light is harnessed in the service of space and time; when it is 'captured' in the photograph.

This perspective is at the core of Virilio's concept of 'dromoscopy', a media and warfare theory, which proposes a science or logic of speed and a study of its impact: 'dromos' is a Greek word for race, as in running, and 'scopy' relates to vision and sight.[57] Virilio states, 'dromoscopy displays inanimate objects as if they were animated by a violent movement.'[58] This transformation of the object through speed and vision has an animating effect on the inanimate through space and time. Virilio's 'dromoscopy' emphasises the dominance of time over space, 'where the idea of nature from the era of the Enlightenment fades away, along with the idea of the real, in the era of the speed of light.'[59] According to Virilio all we have is light, a dromoscopic scenario, where only illusory and fleeting appearances of the dimensions of space and the real come to the fore.[60] As Virilio argues, 'images have become a new form of light, but it's a question of lights that we can't yet understand since we're still attracted by their specularity.'[61] By this, Virilio means that an analysis of photomedia can only be undertaken by detaching ourselves from the seductive glare of these luminous images. To better understand this historical account of 'dromoscopy' or the 'industrialisation of vision', we must first seek to understand photography (which is, at its very core, light) through an analysis of the light images produced by the early inventors of photography.

The photographs by Niépce and Daguerre work to frame the world they found themselves experiencing. They pointed their cameras into a selective 'view' of their immediate environment. Virilio locates the importance of a 'view' for an understanding of things when he notes that 'we live in a world where everything is always co-present, a world without determinisms, where before and afterwards no longer make any sense, and where our gaze, the very fact of observation is what creates effects.'[62]

Niépce's early image frames the world within a single perspective or point of view; it also represents the ability of photomedia to capture space through light. The light-space of *View*... is with us today: we too can gaze out of Niépce's window as he had done. Ann Thomas points out, 'the photographic image becomes the permanent record of the act of observation'[63] and the act of recalling that observation. Like the perception of light itself, these images travel to us from the past and from a 'point of view'.

Another example of framing vision from this era can be found in Talbot's *Latticed Window* (fig. 1.3).[64] Here a ghostly image, obtained from the interior of Lacock Abbey, shows the oriel window at the south wing of the Talbot family home. The printing process Talbot developed allowed a photographic negative, not a positive, to be produced. In this image, the interior, marked by the distinctive latticed window design, is white and frames the exterior view of the sky which is inversely rendered black. Attached to Talbot's notebook the accompanying annotation reads, 'When first made, the squares of glass about 200 in number could be counted, with help of a lens.'[65]

Talbot treasured realism, positioning it as a defining aspect of the success of this image. This is apparent through the annotation attached to the image describing the ability to count all panes of glass on the window, an indicator of the clarity of the photograph. Talbot's attitude, indicative of the drive for realism, formed part of the impetus for the invention of photomedia in which the real is experienced through the image. It also exposes a 'point of view' both in terms of the actual photograph and the photographer's stance on photography.

While Virilio sees light through speed, Baudrillard argues that society has experienced a 'precession of simulacra' which occurs through three successive 'phases' in the history of the image.[66] The first phase, which belongs to the pre-modern era in which images

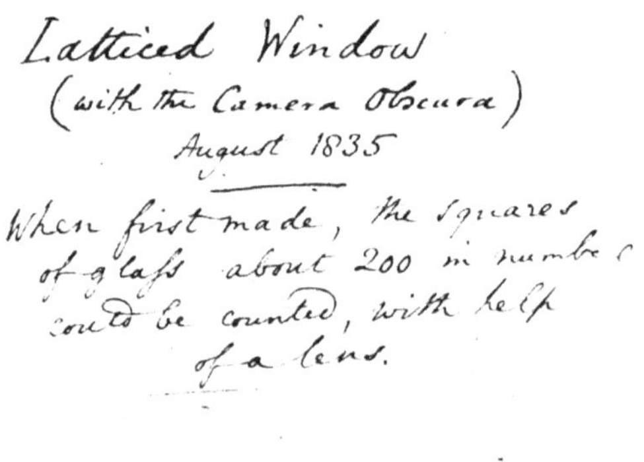
Latticed Window
(with the Camera Obscura)
August 1835

When first made, the squares
of glass about 200 in number
could be counted, with help
of a lens.

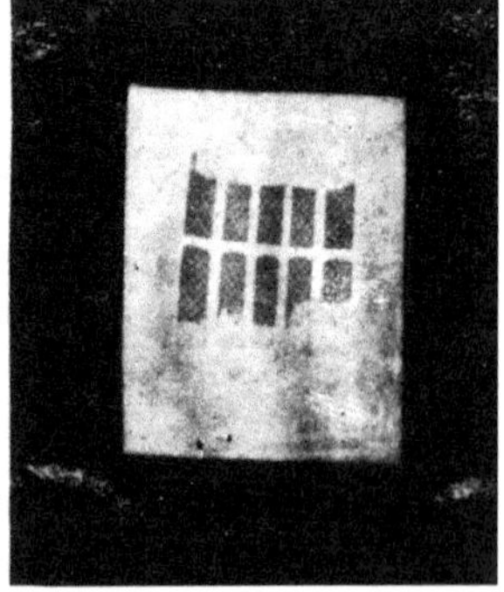

Figure 1.3 William Henry Fox Talbot, *Latticed Window*, 1835.

such as paintings and drawings were clearly copies of an original, preceded the era of the early image machines. In Baudrillard's second phase, in the era of the early image machines which arose with the industrial revolution, the distinction between the image and the real breaks down in photography and other mass media technologies. No longer simply a reflection of reality, the image obscures and displaces the real through photographic representations of it and through the copy the real is not accessible. Baudrillard asserts, 'it is perhaps not a surprise that photography developed as a technological medium in the industrial age, when reality started to disappear. It is even perhaps the disappearance of reality that triggered this technical form.'[67] What actually disappears, or begins to disappear during the industrial revolution, is the notion of a singular reality or one point of view. Photography proliferates and spreads multiple points of view. Baudrillard's claim is startling and has direct implications for the structures of light-space-time. Instead of transporting impressions of space and time, Baudrillard suggests that photomedia transposes them. In other words, the relative position between the image and that which it represents changes sequence so that the image has a greater reality status than reality itself. For Baudrillard, representation is lost in favour of simulation. Baudrillard directs us back to Plato and specifically to the Platonic line of reasoning that, 'the image stands at the junction of a light which comes from the object and another which comes from the gaze.'[68] Both the viewer and that which is viewed form what is seen.

However, in the age of the early image machines, the photomedia image has not yet completed a Baudrillardian succession to simulacrum, it is just beginning. At this stage, for Baudrillard, images such as *Latticed Window* serve to both reflect and mask reality. Talbot's image is intended to provide a faithful representation of his window, and for Talbot, it does. It reflects what is real to Talbot but for us it masks the real. Even if we have never been to Lacock Abbey we have seen and now have knowledge of that window as a result of looking at Talbot's image. In this sense, the image 'stands in' for reality, for as Baudrillard claims, 'every photographed object is simply the trace left behind by the disappearance of everything else.'[69] Certainly, that light-space-time which Talbot photographed is only present in his photograph as an image.

Image-space

This use of photography to capture light-space-time contributed to the manifestation of a new expression of light-space: the image-space. As the era of the early image machines evolved, we can begin to see the enduring processes and structures inherent in all photomedia images as light-space-time based. We also start to see the proliferation and dissemination of photo-images: they become ubiquitous. As a result, image-worlds begin to emanate from the ceaseless production and dissemination of images and their constant reception. This cyclical mechanism of production informing reception, cognition and understanding, which then informs new production and new reception, ultimately develops an inner-space, a personal relationship with images based on one's understanding of them and the inner subjective representations they evoke. This prevailing inner-space influences and is deeply connected to an outer-space. Outer-space is the contrapoint of inner-space in that it is a shared, external space, positioned in the public domain, where photomedia images are disseminated. In this way, image-space simultaneously occupies two spaces: an outer public space and an inner, subjective space.

Hippolyte Bayard's *Self Portrait as a Drowned Man* (1840) is an early example of the connection of inner- and outer-space through the image-space of photomedia. The transference of personal narrative from inner-space into outer-space is shown in Bayard's image of the photographer, himself, as a drowned man. A white drape falls with classical folds over Bayard's torso preserving his modesty, a cane hat hangs to the left of centre and a vase sits on the right side of the image. Bayard's limp body is propped up, slumping slightly to the right; his face and hands appear black, a visual indication of his corporeal decay.

The degraded character of the image, a result of the printing technique developed by Bayard, imparts an atmospheric quality. One can envision with equal measure the scene in which Bayard is pulled from the murky depths of the lake, already dead, and the theatrical melodrama of Bayard as he enacted the part of a drowned man. A smirk might be anticipated on the sullen mouth of Bayard at the humour of 'fooling' the viewer; however, this is not the case. Instead, Bayard plays the scene with deadpan expression. On the verso of this image, he penned a third-person account of his fictional suicide.

Geoffrey Batchen's analysis of this image, in his book *Burning with Desire*,[70] draws parallels between *Self Portrait as a Drowned Man* and Jaques-Louis David's *Death of Marat* (1793). Batchen claims that Bayard deliberately referenced the image to make a subversive political statement during what was a turbulent time for French politics.[71] Batchen claims that the image was a comment on the competence of the government and the legitimacy of the King.[72] This may be the case, but Batchen also acknowledges that *Self Portrait as a Drowned Man* speaks of larger issues attesting that the image is a response to the acknowledgement by the French Academie des Sciences of Louis Daguerre as the inventor of photography. Despite Bayard's advancements toward the technology, he failed to gain acknowledgement for his achievements and the sense of his loss is palpable within this image.[73] *Self Portrait as a Drowned Man* eloquently speaks of the outside forces that resulted in Bayard's metaphoric death as Daguerre enjoyed a place in history while Bayard was left to fade into obscurity. However, Bayard is not lost in history; this single image positions him as a visionary. It exhibits a profound realisation of the ability of photography to capture a moment that exists in inner-space, re-position it within outer-space and thereby circulate in image-space. *Self Portrait as a Drowned Man* demonstrates the cyclical nature of inner and outer-space coalescing within image-space as the unseen becomes seen. In Bayard's photograph, inner comes to form outer; however, its very presence within the world contributes further to the cyclical flow between inner and outer-space.

As Gaston Bachelard maintains in his book *The Poetics of Space*, 'outside and inside form a dialectic division, the obvious geometry of which blinds us as soon as we bring into play metaphorical dimensions. It is the sharpness of the dialectics of yes and no which decides everything.'[74] Inner-space – the internal, personal and perceptual space of subjective experience – and outer-space – the external, public space, the impersonal space we all inhabit – are brought in direct correspondence with each other through photomedia. As Baudrillard observes, 'The miracle of photography, of its so-called objective image, is that it reveals a radically non-objective world.'[75] It is in fact a representation of inner-space and outer-space, which forms the dynamic, evolving structure of the image-world.

Seeing machines

At this early stage of photography the camera was not only used to capture the otherwise imperceptible realms of inner-space, it was also used to capture that which lay beyond normal perceptual boundaries; that which went beyond immediate humanly perceivable physical space to investigations of micro and macro spaces.

Photography and microscopes were brought together as early as 1835 through Talbot's experiments with his solar microscope and salted paper.[76] Once photography was more clearly established and made widely available, many other scientists used it in combination with microscopy and the conjunction of these technologies became known as photomicrographs. The appeal of this process was that the invisible, interior space of substances was effectively transcribed and made visible with an accuracy and objectivity important to scientific analysis.

Neuroanatomist Dr Jules Bernard Luys[77] chose the process of photomicrographs to illustrate his studies of the cerebellum in 1873.[78] He states that in using photomicrography he was 'substituting the action of light for [my] own personality, in order to obtain an image [that is] both impersonal and accurate'.[79] With this statement, Dr Luys acknowledges not only the ability for photography to bring the micro into the macro for clarity of visual perception but also that the image is veridical.

Scientists and astronomers used the new technology to speedily capture what they saw through their microscope and telescope lenses. The astronomer John Herschel[80] was interested in what photography might tell him about light most likely to aid his own observations of the cosmos. Herschel also made several important contributions to photography such as the cyanotype process and published several articles on the effect of light on photographic materials.[81] Also, he coined the word 'snapshot' in 1860 referring to the fast capture of light by the camera.[82] Another astronomer interested in capturing images of distant spaces was John Adams Whipple, the first person to successfully take pictures of the Moon and stars using a telescope and a camera. *The Moon* (fig. 1.4) was taken using the Daguerreotype process. The image clearly depicts the lunar surface marked by the impact of asteroids and comets. Within the image, the spherical shape of the Moon is emphasised by the strong side lighting of the sun and the faint presence of stars at the left side of the image. The Daguerreotype process used by Whipple endows the image with a unique quality. The process allows just one positive image to be made, through the exposure of a polished silver mirror surface coated in silver halide particles deposited by iodine, bromide or chlorine vapours. The luminosity of the Daguerreotype is profound; a silvery mirror of light forms the substrate of the image. The result is an image that engages light in a reflexive manner. Both the light-space of the image and that

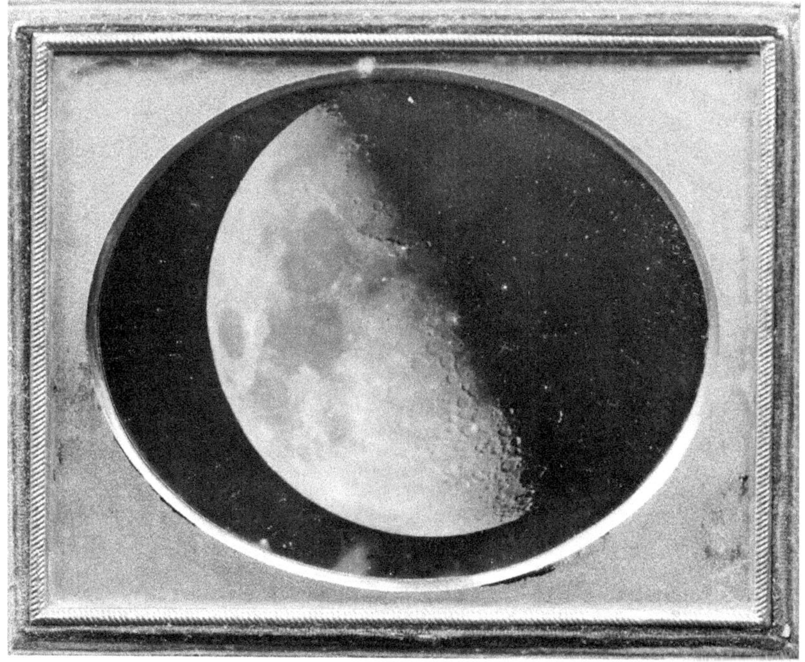

Figure 1.4 John Adams Whipple, *The Moon*, c.1840s.

of the space in which the Daguerreotype is viewed can be seen, the image and the viewer may also be seen in the same object. This effect is produced by the mirrored surface of the Daguerreotype, causing the photographic image and that of the viewer to come together.[83]

Light is used by Whipple to capture space, forming a light-space impression of the Moon. Similar to Bayard's *Self Portrait as a Drowned Man*, this image, *The Moon*, locates an object which is normally outside physical proximity within the here and now. While in *Self Portrait as a Drowned Man* the inner-space of imagined reality is brought to the fore within the image-space of the photograph, in *The Moon* it is the physical universe which lay beyond normal perception that is brought close. This photographic act renders the Moon with the same apparent qualities as anything else that is photographed; through the image, the light-space structures are given a presence. *The Moon* is a demonstration of the desire to bring physically distant objects visually close, locally fixed in time and space through photomedia.

This cataloguing of physical reality through photography resulted in a perceptual contraction of geophysical space. The photograph brings the moment and objects proximately closer. It places 'there' closer to 'here'. As Virilio maintains, this is the beginning of the transformation of society through 'dromoscopy'. Virilio reasons that the developmental axis of this new image-space begins with the invention of photomedia, as this is the inauguration of this type of space: light-space. As Virilio claims, with photomedia the 'transportation evolution' begins with the introduction of a new space which is light based[84] in an era which was conscious of the first jolts of speed. What was felt was the beginning of an exponentially increasing speed, and nineteenth century society felt this somewhat like a steam engine leaving the station. We can see this in the work of the early photographers; their images maintain the unmistakable mark of speed: the blur. The blur is suggestive of the era that photography was born into and of the technology itself, its development and movement forward.

Paul Virilio speaks of 'visual retention' as the ability of the visual to extend beyond the retina. He explains that we receive light and that it travels as electrical impulses to our nervous system, meaning that within the body light has a molecular basis.[85] In *The Vision Machine*[86] Virilio cites the writings of Disidéri[87] who, like Daguerre, was driven to speed up the process of photography with his visionary goal of obtaining an instantaneous production of the image. As Virilio maintains, 'seeing the world becomes not only a matter of spatial distance but also of the *time-distance* to be eliminated: a matter of speed, of acceleration or deceleration.'[88] Virilio proposes that spatial distance had been a dominant paradigm and through photomedia, time and speed became part of the visual understanding of the world. The experience of time was altered with the advent of photography. Later cinema would also extend these light-time effects. But, in the era of the early image machine, before the coveted 'instantaneous' moment could be grasped, we see photography bring about new relationships to light-time. In many other ways this epoch is marked by accelerating speed but in terms of the history of photomedia this is an era of slow time, a time when luminous duration comes impressed upon the photographic surface, bearing down with an intense physicality.

The first photographic images are testimony to the relationship photography maintained with light-space-time since the beginning of the early image machines. From 1839 to 1880 the technology of these image machines was continually advanced. It was a time in which the industrial revolution progressed with an incredible urgency. Society accelerated with the speed machines of industry; as technologies such as steam and air travel advanced, so did the transfer of information with the telegraph. A few years after *View*...was taken, Niépce had increased the light sensitivity of the photographic surface so that around 30 minutes was all that he required for recording an image. However, that was just the beginning. The photographic process, of capturing light, continued to accelerate and exposure times reduced dramatically. The first aerial photographs taken by Félix Nadar[89] in 1862 had exposure times of 20 seconds.[90] Immediately popular, photography was used to 'capture' everything and with increasing speed. Overall, the technology gained greater functionality, with the most notable improvements occurring through changes in light sensitive materials and image development processes.

The early photographers such as Niépce, Talbot, Daguerre, Bayard and Whipple established the fundamental properties of photomedia. Light, the primary material of photography, contracted space and time, bringing objects and moments closer through luminous and weighted stillness. It also framed the view of those who used it, whether that was the view of the external (physical) world or the internal (psychological) world. Photography provided a powerful technological tool and the effects were profound from the earliest image machines onwards.

While these early image machines were seen as an invention to serve humanity, the German philosopher Vilém Flusser took a different view when considering the nature of photography in his book *Towards a Philosophy of Photography*.[91] Flusser directs the reader toward a central premise, based on two key developmental points in human culture. The first is the invention of linear writing and the second is the invention of technical images. Technical images are defined by Flusser as technically or mechanically produced images that have been created by an 'apparatus' (a camera or computer). His theory that the apparatus functions to program society for the improvement of itself develops in several key ways. Flusser ascribes the drive for the technological progression of the apparatus to an inherent property of the technology, which is that we improve the technology because we are part of the program of the apparatus. Flusser considers the apparatus to function not in the service of human communication but in the service of itself. He positions the locus of control with the apparatus, and says that the apparatus is not only in service of itself but also in control of human civilisation. Another aspect of this program is the proliferation of 'technical' images, which form an image-space or as Flusser claims a 'photographic universe'. Flusser states that 'The invention of the photograph is a historical event as equally decisive as the invention of writing.'[92] Flusser's theory highlights the importance of the invention of photography and its connection with language, or more precisely with messages and understanding. Pertinently, Flusser separates the invention of photography from writing; although both are communication devices Flusser places photography in its own sphere, claiming it as a new language.

Ever brighter: moving on from the glow of the early image machines

From a contemporary analysis of the history of photomedia it is clear that this era of image machines heralded the initial formations of light-space-time structures. A fundamentally light-based medium, the relationship of the photographic image to light is at once physical, psychological and philosophical. This relationship has the power to form and transform our understanding of space and time. The slow exposure time of the early image machines, most notable in Niépce's *View…*, was contained within the photographic image and was impressed upon the viewer. From the earliest conception of photography, space contracted through the image as objects were brought perceivably closer. Through these light-spaces new internal territories such as Bayard's *Self Portrait as a Drowned Man* both initiated and contributed to the formation of image-spaces. As the technology progressed in association with its increased application so did these light-based structures. This was a new era with a new light-based image: the photograph.

The early photographs, made by the artists and scientists who worked to develop photography, are indicators of new experiential and reflective spaces. These images provided a new relationship to light and subsequently altered relationships to space and time. This effect is unparalleled in history and one which continues today in photomedia technology. The possibilities for art and experience altered with the invention of photography. Used to reveal places and spaces beyond the capability of the eye, worlds within worlds, alternate worlds and distant lands were made visible through the mechanisms of photography. The vast volume of space was compressed and transported through time.

Light and time formed a symbiotic relationship and conveyed these new spatial configurations. The broader field of photomedia, which began with photography, maintains a potent relationship with space and time through light. Photography does not open up a direct relationship to the space that exists in front of the aperture as Niépce, Daguerre and Talbot believed. Rather it becomes a fragment of a larger perspective that exists alongside many other views and forms a new image-world, new image-spaces and a new language. Intimations of these characteristics were already present in the earliest images. This becomes highly apparent during the next era of the analogue image machines.

2

ANALOGUE IMAGE MACHINES C.1870–C.1990

Super vision: the analogue era

Photomedia, growing from roots firmly established in the era of the early image machines, experienced unprecedented expansion in the analogue age.[1] The fundamental stages of photographic technology had been refined by the 1870s, providing the basis for rapid development of analogue technology which continued for over a hundred years into the 1980s. Unlike any other historical period before it, the analogue period was a time of profound change across all sectors of society. It was an era which began with train travel and ended with space travel. The same forces that propelled the epoch forward also advanced photomedia exponentially. Many important photographic progressions occurred in the analogue era including shutter technology, extremely light sensitive film, flash lighting methods, colour images, durable plastic film, Polaroid systems, cinema, television and video to name just a few. More generally, these advancements resulted in greatly improved and simultaneously miniaturised camera body technology and methods for recording light. Concurrently, mechanisms for displaying photomedia evolved to encompass a wide range of printing and screening devices. These analogue methods for the capture, storage and display of light vastly increased the scope, precision and visibility of photomedia.

However, regardless of these technological advancements, light, the fundamental property of photomedia, remained constant in the analogue era. The analogue image machines strengthened, evolved and altered the light-space-time structures that first glimmered with the early image machines. In the analogue era light-time gained speed, surpassing the slow time of the early image machines. Slow time transmogrified into fast time, which came to be indicative of the era. Light-space continued to contract geophysical space and proliferate an image-space which propelled photomedia forward and contributed to its subsequent ubiquity.

Standing still: the instantaneous capture of light-time

The instantaneous capture of light had been a goal of Daguerre's at the time of the invention of photography. By 1873 the photographic apparatus was capable of recording events at 1/1000 of a second, essentially achieving Daguerre's goal. Unfortunately, he was no longer alive to witness it. Central to the realisation of instantaneous photography were the developments of sensitive photographic plates and Ottomar Anschütz's focal-plane shutter, the device that enabled shutter speeds close to 1/1000 of a second.[2] No longer bound to the lengthy exposure times of the early image machines which registered a blur with any movement, these new analogue machines seized light at a rapid pace, rendering a moment in time with crisp clarity. With instantaneous photography, actions that moved faster than the human eye could physically see were rendered immobile in an instant and made discernable through the photograph. Immediately, the technology was applied to scientific fields for the examination of physical locomotion by the likes of Eadweard Muybridge,[3] Étienne-Jules Marey,[4] Albert Londe[5] and A.M. (Arthur Mason) Worthington[6] in the nineteenth century; and by Harold Edgerton in the twentieth century.[7] As discussed in Chapter 1 through analysis of Joseph Nicéphore Niépce's *View...*, c.1826 and Louis Jacques Mandé Daguerre's *Boulevard du Temple, Paris,* c.1838, photomedia introduced light-time structures which contributed to a new perspective concerning the subjective experiences of time. In the analogue era, images such as Eadweard Muybridge's *The Horse in Motion,* 1878, Étienne-Jules Marey's *Movements in pole vaulting,* c.1900 or Harold Edgerton's *Bullet through Banana,* 1964, display the distinctive characteristics of instantaneous photography and the ensuing time and space structures as they formed in this era.

In Eadweard Muybridge's *The Horse in Motion* (fig. 2.1) it is possible to see beyond the normal visual threshold as the flow of time lingers for optical inspection. As Ann Thomas notes, 'we can never see these moments by means of direct observation.'[8] Instead, we see them mediated through the light capture made possible by advances in photomedia technology.

Artists had for centuries debated the true position of a horse's gait, some believed that two hooves stayed firmly earthbound at any one time and illustrated this in their work, while others postulated that all four hooves left the ground and represented this accordingly. In 1881 Eadweard Muybridge, while on a European information tour, delivered revelatory lectures with the aid of *The Horse in Motion* and other chronophotographs he had taken.[9] These images revealed with visual clarity that all four hooves left the ground mid-gallop. From this moment on, artists represented this in their work. *The Horse in Motion* is a clear example of the capability of photomedia to alter perception.

Muybridge achieved his chronophotographic study by arranging a series of 12 cameras and triggers, which were sequentially released as the horse and rider moved along a set path.[10] The individual photographs were later compiled into what appears to be the chronological order of events to convey the progression of movement. Each image maintains a temporal position relative to the image immediately before and after it. If the intended visual chronology is followed by the viewer a sense of continuity is obtained. However, the chronology is easily broken because each image is presented as a separate and discreet moment composed in a grid formation; each movement appears as a frozen fragment of time. The intended sense of continuity is only properly achieved when the frames are scanned by the eye in quick succession and even then Muybridge's chronophotograph retains a static quality. Before and after are lost in favour of one moment or the next. As Thierry de Duve affirms, Muybridge 'demonstrated what the animals' movements were, but did not convey the sensation of their motion'.[11] This is probably the reason Muybridge re-animated his chronophotographs through a Zoopraxiscope.[12]

Temporal reconstruction was an important aspect of Muybridge's oeuvre. In several of his chronophotographs moments

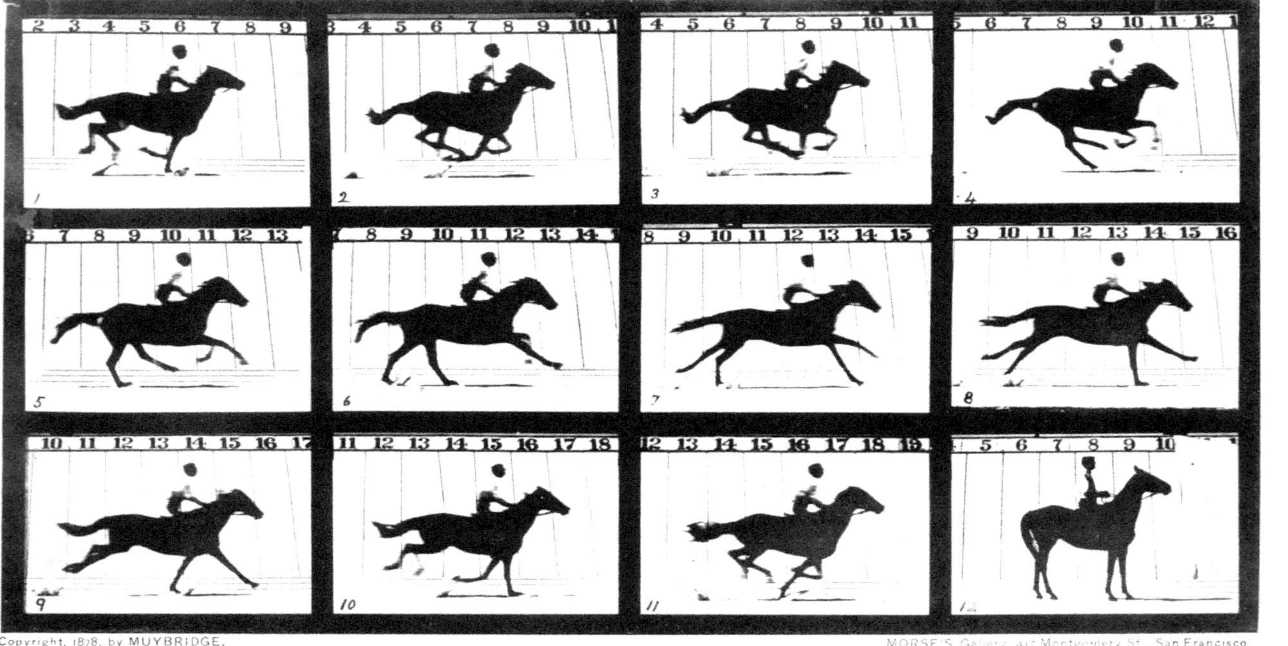

Figure 2.1 Eadweard Muybridge, *The Horse in Motion*, 1878.

have been cleaved from the continuity of time and reconfigured out of succession; in some instances images are repeated, omitted or inserted to increase visual impact and supply a prescribed narrative. An example of this temporal reconstruction can be seen in *The Horse in Motion*. Note that in these images the horse is in full gallop from the first frame, resting abruptly by the last. Muybridge intended to employ a linear, Cartesian and scientific methodology as a way to 'guarantee its scientific accuracy'.[13] However, his pseudo-scientific approach of segregating and (re)configuring light-space-time mainly directs the viewer to a perspective that is equally fragmented and relative to our own position in space and time. *Horse in Motion* provides a non-linear progression of separated stills, reconfigured but not recombined. Tom Gunning writes on Muybridge, 'his discoveries rely upon a modern rational and scientific mastery of space and time in which each has been calculated and charted in relation to each other.'[14] For this reason, they are a remarkable if unintentional connection to the relative nature of space and time that Albert Einstein would espouse over 30 years later through his *Special Theory of Relativity* (1905).[15] This theory, which opposed the Cartesian model of space-time, posited that relative to the observer, both space and time are altered when approaching the speed of light. If one could possibly travel at this speed, space would stretch out and time would move so slowly it would be perceived to be still.

Muybridge, having received a positive response to his early photographic investigations into motion, wanted to capitalise on a public fascination for his images. An opportunity presented itself when William Pepper,[16] a professor of clinical medicine at the University of Pennsylvania in Philadelphia, USA, convened a committee in 1883 to subsidise further photographic studies of animal and human movement.[17] Within a few years Muybridge had made almost twenty thousand single images including *Man Walking Downstairs* (1884) and *Woman Walking Downstairs* (1887). These chronophotographs which show nude male and female figures descending a staircase directly influenced another chronophotographer, Albert Londe, to undertake a similar study.[18] Marcel Duchamp's painting *Nude Descending a Staircase, No.2* is clearly inspired by his knowledge of chronophotographic studies.[19] However, unlike Muybridge, Duchamp was not only concerned with physical movement but also with the representation of physical and psychological transformation through time and space. These he explored by using a Cubist method of fracturing the subject of his work into several planes, which Duchamp himself labelled 'dismultiplication'.[20]

Muybridge's vast exploration of movement through chronophotography is more than scientific explorations of physical locomotion; the photographs are voyeuristic. On closer inspection of Muybridge's larger oeuvre it is apparent he was highly attuned to the idea of spectacle. Many of Muybridge's chronophotographs show nude or semi-nude human bodies – women, children, strong men, the disabled and disfigured – demonstrating various characteristics through a range of movements. Further contributing to the spectacle of his chronophotographs, Muybridge would reanimate them via the Zoopraxiscope. The complete body of work produced by Muybridge earned him the title 'father of the motion picture' by some.[21]

Like Muybridge, Étienne-Jules Marey's photographs reveal space and time beyond the threshold of human ocular capabilities. In Marey's images, gaps and missing pieces, like the blank spaces between film frames, are present within a single image. However, in contrast to Muybridge's work one moment flows into the next and the entire set of physical actions persists visually over time, and are contained within a single image, somewhat closer to Duchamp's *Nude Descending a Staircase*. This has the effect of collapsing time into a moment in light-space.

To obtain images such as *Movements in pole vaulting* (fig. 2.2) Marey developed a unique photographic system that involved isolating the subject on a dark background. Marey also organised a photographic studio which he transformed into a black space. These black spaces were often many metres deep and tens of metres long, allowing Marey to capture the full range of movements against a dense, black background. At times Marey had his subjects dress in a black body suit with reflective white points connected by reflective lines to denote aspects of the body. Through this method Marey abstracted human movement by separating the line and trajectory of the physical body from extraneous visual information. While the subject was in

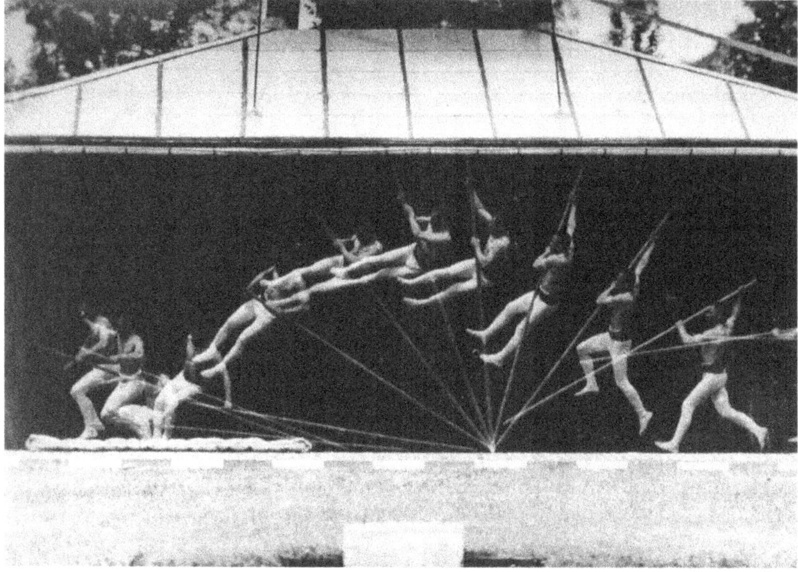

Figure 2.2 Étienne-Jules Marey, *Movements in pole vaulting*, c.1900.

movement the aperture of Marey's camera was left open. Attached to the lens was a rotating metal disc with one to ten slots cut into it at equal distance. As the subject moved, his actions were recorded on a different part of a glass plate, in sequence.[22] The resulting image demonstrates physical movement through space and time without requiring a bank of cameras or photographic plates. Marey devised other methods for obtaining his images. To photograph birds he assembled a photographic gun; similar in shape to a rifle, it was able to shoot 12 frames a second on a disk.[23] In all of his photographic work Marey aimed to 'combine visual evidence with a precise registration of the passage of time'.[24] Before working with photography he was a physiologist and researched the circulation of blood, for which he devised specialist instruments. This same scientific approach was applied to his photographic endeavours, producing images that aided understanding of otherwise unseen natural phenomena. For instance Marey's publication *The Flight of Birds*, which features images 'shot' with the photographic gun, was used in aviation studies and resulted in an understanding of lift and drag forces essential to the development of the aeroplane.[25] Marey's investigations were so revealing that they assisted others with enquiries concerning motion pictures and athletic training.[26]

As Cartwright points out, 'Marey effectively reorganized the body to make it embody its own status as an object subject to the laws of temporality and duration.'[27] In *Movements in pole vaulting,* Marey has exposed the fragility of time; each fraction of a second reduced to a paper thin impression. The figure is translucent as it moves from the right side of the image to the left. Past, present and future merge as the gossamer man prepares to jump, levers himself over the hurdle and lands at the same time. The delicacy of that moment, and to a larger extent of time itself, is made visible. Marey, like Muybridge, succeeded in revealing what is not otherwise apparent to the human eye.[28]

This reduction of the body to its fundamental form and movement through space is understood by Derrick Price to contribute to an overall perception of the human body as a machine. Price states that 'this perception depended on a highly idealised and utopian concept of the efficient, smooth-running machine, which ignored the ways in which certain kinds of machines mutilate bodies (for example in factories and war).'[29] Although Price makes a salient point, what is most pertinent here is not the relationship Muybridge's or Marey's work maintains with factory machines or war machines but the connection that images such as *The Horse in Motion* and *Movements in pole vaulting* have to the image machines of the analogue era and the related space-time structures they initiate.

This new instantaneous photography introduced dualistic characteristics to the medium, binding light and time in a profoundly new way. This light-time duplex is snap-seized through the rapid camera action. It persists, motionless, through the continual flow of time. In other words, the fleeting moment, too fast for our eyes to register, is immobilised for later visual reception and in-depth analysis. Virilio claims that, 'through its unrealistic play of visual techniques, its slicing of reality, its immobility, its silence, and its phenomenological reduction of movements, photography affirms itself as both the purest and the most artificial exposition of the image.'[30] Time is simultaneously exposed and artificially stopped through chronophotography.

Other contributors to the field included Albert Londe, Thomas Eakins and A.M. Worthington. Worthington produced *A ball falling into a mixture of water and milk and the ring of turbulence thus produced* (1894), a precursory image to the work of Harold Edgerton's famous image of a milk drop, *Milk drop coronet* (c.1936) which shows the precise nature of a drop of milk at the moment of impact with a hard surface.

The practitioner Harold Edgerton developed his own system that was similar to Marey's; however instead of using a disk with slits to expose the negative to light Edgerton simply opened the shutter for a fraction of a second and continuously ran film through the camera while high-speed bursts of light were emitted from a stroboscope.[31] From the 1930s onward, Edgerton continued his specific line of scientific enquiry for decades. Unlike other chronophotographers' studies, Edgerton at times selected a single image to represent the event, such as the explicitly titled *Bullet through Banana*, an image that is indicative of Edgerton's larger body of work. From a drop of milk to a banana gushing from a bullet wound, Edgerton's images used instantaneous photography to expose discreet moments of time as frozen and quantified forms. The results were concurrently prosaic and unimaginable, a common banana and a normal bullet suspended in a surreal moment.[32]

With instantaneous photography Muybridge, Marey and later Edgerton participated in the capture and configuration of light-space-time. With their images they depicted the normally unseeable, and succeeded in expanding and simultaneously halting light-time.

This fascination for stopping and capturing minute quanta of time allowed Muybridge to produce images that halted and broke up the moment. Marey portrayed a similar treatment of time while allowing a delicacy to materialise as a sequence of moments played out within a single photographic frame. As Marta Braun points out, 'they provided an iconography of dynamism for the art of the twentieth century, and helped change forever our experience of time and space.'[33] Edgerton conversely succeeded in suspending time at a surreal and sculptural point. Their work provides a profound insight into the precise moments of light-time while irrevocably altering our understanding of time.

Baudrillard contends that 'the stupefying power of the photo is far superior to that of writing. It is rare that a text can offer the same instantaneity, the same tangibility, the same magic, as a photographic object (shadow, light or material).'[34] Certainly, it is unarguable that what occurs in these images is specific to photomedia.

Moving quickly: Photofuturism and light-time

There was another response to breaking down light-space-time into structured and direct lines of movement as explored in chronophotographic studies, which can be seen in work by participants of the Photofuturist movement. This movement emerged in tandem with the Futurist art movement of the early twentieth century. The self-proclaimed Italian leader of Photofuturism, Anton Giulio Bragaglia (1890–1960), penned the manifesto titled *Futurist Photodynamism* in 1911.[35] In keeping with the many other Futurist manifestos, the Photofuturists set out a proposition to assert the free and expressive sensation of speed and movement through their art.[36] In opposition to chronophotography they sought to create non-representational images exhibiting characteristics that were unclear, indiscernible, confusing even. Their work aimed to portray a pure expression of the sensation of movement indicative of the new industrialised society. Bragaglia proclaimed that 'We are not interested in the precise reconstruction of movement, which has already been broken up and analysed. We are involved only in the area of movement which produces sensation, the memory of which still palpates in our awareness.'[37]

This approach allowed an elastic, temporal space of free flowing movement to form within the picture plane where beginning and end-points of actions blend and are difficult to discern. Importantly, the Photofuturists depicted that which cannot normally be seen, their psychological view, as opposed to the chronophotographers who photographed that which cannot be seen directly through physiological means. The Photofuturists aimed to create images that engaged with the psychological realms of speed and movement; to represent this they 'attempted to render movement by vaporising form'.[38]

In Bragaglia's image *The typist* (fig. 2.3), the movement of a typist's fingers are combined as one action on a single photographic plate. Unlike Muybridge's *The Horse in Motion* or Marey's *Movements in pole vaulting* there are no structured gaps in the physical movement of the subject. While Muybridge's and Marey's work reduced form into its kinetic phases to demonstrate physical movement through time and space, *The typist* shows a new treatment of movement and time.[39] In *The typist* physical line and trajectory are merged into a fluid continuum, the beginning and end points are indistinguishable as they flow through space and time. In this image a large black typewriter sits unmoving and firm on the typist's desk, her pallid hands endowed with a spectral quality because of their fast movement during the exposure of the image. Here the typist is fluid and the machine is unmoving, a distinct metaphor for the adaptation of the human body to mechanical processes during the early twentieth century. Also, while viewing this image one can assume the intended perspective of a stationary viewer observing a world which rushes by in a blur. Past, present and future merge, blurring movements and moments within a compressed light-space-time,

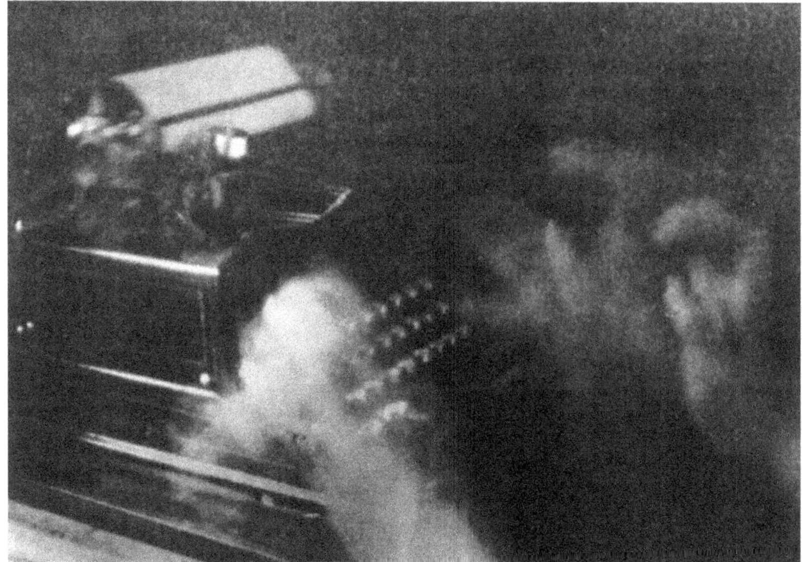

Figure 2.3 Anton Giulio Bragaglia, *The typist*, 1911.

forming a sensation rather than a pictorially discernable or clear image. Bragaglia reasons, 'the movement of light [is] revealed in Photodynamism.'[40]

The typist illustrates, through the photographic capture of light, a consciousness of the era's space and time structures that others such as Wassily Kandinsky considered years later. Kandinsky remarks that, 'since 1914, the tempo of time seems to have been increasing in speed. Inner tensions accelerate this tempo in all spheres familiar to us. One year is possibly the equivalent of at least ten years of a "quiet", "normal" period.'[41]

The typist is expressive of a larger sensation of the epoch. As Paul Virilio writes, 'they saw every vehicle or technical vector as an idea, as a vision of the universe, more than its image.'[42] The Futurists and the Photofuturists understood the progression of twentieth-century technological advances as a process of breaking away from old values and hierarchies.[43] In the constant drive to explore their vision the Futurists took up the new technology of cinema.[44]

Cinematic light-time

Futurist cinema, like the Futurist art movement in general, was interested in producing non-objective methods of expression. The visual effect of Futurist cinema was intended to jolt the senses, create a feeling of agitation and dynamism through movement. This was achieved through new approaches to cutting between frames, scratching and painting on the surface of the film, and staging highly aestheticised and choreographed sets and props which, through their geometric composition, cut through the picture plane. Futurist cinema was not the only attempt to signify new perceptual realities. In this era Fernand Léger's *Ballet Mécanique*, 1924, Marcel Duchamp's *Anémic Cinéma*, 1926 and Hans Richter's *Rhythm 23*, 1923 are but a few other examples from the early part of the era.

One of the most influential artists working with photomedia at this time was László Moholy-Nagy, who in the twentieth century declared that 'this century belongs to light.'[45] He opposed the notion of purely representational photomedia and instead considered it to be a 'light form'.[46] Moholy-Nagy's preoccupation with the phenomenon of light and its inherent power to give form led to explorations in camera-free photograms, which he expounded as 'writing with light'.[47] These light-forms were essentially utopian; they expressed new ways of seeing in an attempt to reconcile art and humanity with the machines of the technological age.[48] Moholy-Nagy's theories and experiments culminated in his film *Lichtspiel Schwarz-Weiss-Grau*[49] which was made in conjunction with the artist's sculptural creation, the *Lichtre-quisit*[50] (fig. 2.4), a kinetic structure made of glass, metal, wood and plastic which reflected available light, mirroring and multiplying it to form a light-play.

In *Lichtspiel Schwarz-Weiss-Grau* a constant mechanical rhythm of light, space and time pulses and flows during its six minutes running time.

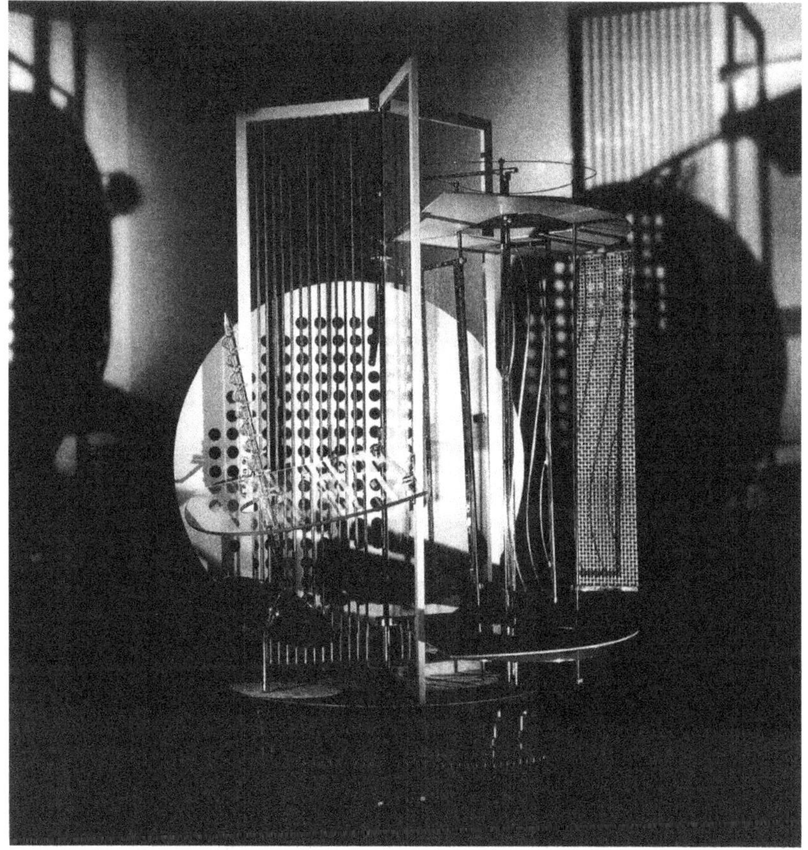

Figure 2.4 László Moholy-Nagy, *Lichtre-quisit*, 1930.

Footage is overlayed, allowing forms and various densities of light to build and compound gradients of light. The film hypnotically flows, permitting a delicate unfolding of time through a steady emanation of light and shadow. The mechanical apparatus set in motion is not only the camera but also the modernist sculpture – a moving, light refracted space – which Maholy-Nagy made the heart of this work. Instead of directing his camera at the world, Maholy-Nagy focused on his light-space construction, a representation of the various elements of space and time he had experienced. The film is suggestive of an encounter one might have at any time of day or night in a gleaming forward-moving metropolis where shadows and light merge, shift and delicately disappear with the rhythm of the city; it merges technology and human experience of the era. *Lichtspiel Schwarz-Weiss-Grau* depicts space and time through its fundamental and reductive elements; it is demonstrative of Moholy-Nagy's utopian vision concerning light.

For Maholy-Nagy light is supreme in the twentieth century but for Virilio it is the speed of this light that is important, claiming that 'the one constant of the twentieth-century was the speed of light.'[51] Capturing light-time accelerated as Virilio attests, 'with photography, seeing the world becomes not only a matter of spatial distance but also of the *time-distance* to be eliminated: a matter of speed, of acceleration or deceleration.'[52]

Still and moving light-time

While Maholy-Nagy explored light through the still and the moving image, others came to explore a new relationship that formed between the still and the moving image with the introduction of cinema. The moving image introduced an impression of events as they happened, and consequently the still image came to function in direct relationship to the moving image. Through the introduction of movement, stillness gains polarity and character. While the moving image offers a temporality, its ability to do so accentuates the inert qualities of the still image. This light-space modulation between the still and the moving image can be seen across the history of photomedia. An early example is Muybridge's chronophotographs such as *The Horse in Motion* (fig. 2.1), which were presented as stills but also animated through a Zoopraxiscope. A later example is Anthony McCall's *Line Describing a Cone* (1973), which used film and light to slowly arc a projected, thin, ray of light into a physically manifested cone of light projected into the installation space.

David Campany locates the invention of cinema with the initiation of stillness, asserting that 'cinema, we could say, was not just the invention of the moving image; it was also the invention of the stillness of photography.'[53] As the moving image manifested in film, and later video depicting the flow of time, the still image commemorated the moment. It is no coincidence that with the invention of moving images, postcards displaying still images became a popular form of visual representation in Europe and North America.[54] The still image created weight and gravity as well as clarity for any event by halting time and thereby isolating the moment; the photograph now functioned with the materiality of an object, a totemic signifier of the event, now past. The still photographic image allows the viewer to reflect on the intricacies of light-space held in place by the fast capture of light-time. This stillness invests the moment with importance and in this the essential power of the photographic still is made manifest. Even when presented as a blur of movement the still image retains its strength, its durability. Those who were aware of this emerging relationship used this new understanding in their work.

Michael Snow, with his seminal film *Wavelength* (fig. 2.5), engaged directly with the light-space-time qualities of photomedia and its still and moving characteristics. The title of the film references the fundamental characteristic of light, which is essentially a particle behaving as a wave, and sound; as Steve Reich explains, 'wavelength – as in length of sound or light wave; wave as in the sea'.[55]

Snow presents a tracking shot over 45 minutes and in the process deconstructs the experience of watching the medium of film. During this time the camera moves from the back of a partially empty room to a point between two windows where there hangs a photograph of a wave.[56] Although this appears to be a single, continuous tracking shot the movement is constructed from footage taken over several different days, on a number of separate film stocks. The accompanying soundtrack consists primarily of a

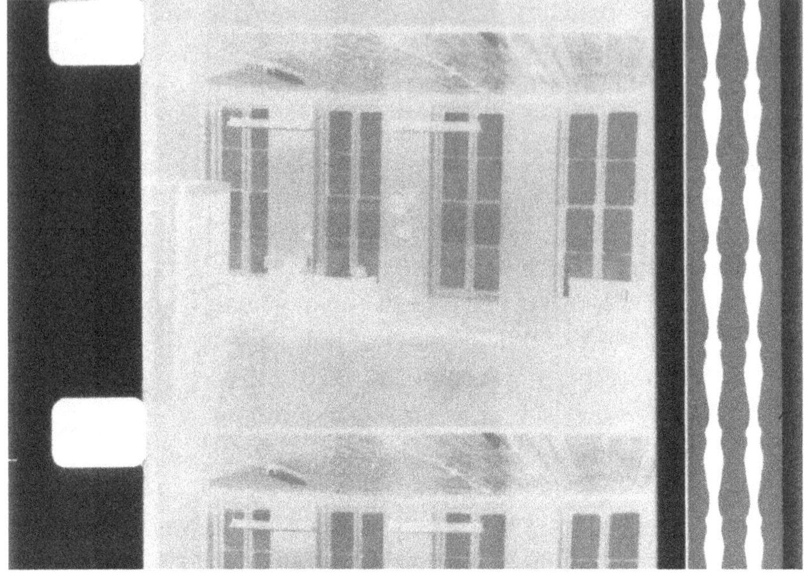

Figure 2.5 Michael Snow, *Wavelength*, 1967.

sine tone (itself a wave), which buzzes with great intensity; as the film progresses the sound falls away at various stages only to increase again at a later point, undulating between the two, again, like a wave. As well, different wavelengths of light penetrate the ocular nerves as an array of duotone images, of various colours, create a 'projected moving light image'.[57] Once on this rolling journey of *Wavelength*, one may experience a sensation of the dilation of time similar to that experienced when moving at immense speed. The primary action within the film involves the movement of the camera as it tracks and the occasional appearance of people entering the field of view, talking on a telephone which hangs at the far end of the room, and then leaving the shot. Watching the film can be at times difficult, as one is made aware of one's corporeal self through the sensation of seeing and listening intently. No reprieve is given through narrative or pictorial clarity. The total experience of light and sound allows sensations of the physical act of seeing or listening to come to the fore: acts which are often forgotten when passively engaging with the moving image.

Virilio compares the sensation of the speed of movement to that of the cinema where the *voyeur-voyager* is like the spectator in the cinema. 'It is he who is projected, playing the role of both actor and spectator of the drama of the projection in the moment of the trajectory, his own end.'[58] The connection between the speed of light-time and the image-space of cinema is profoundly expressed in *Wavelength* through an examination of the materiality of the moving and the still image, as well as the essential properties of light and sound, as waves. Furthermore, like Virilio, Snow signals an affiliation between the experience of the spectator and death when, during the film, a person enters the room, appears to perish as they fall to the ground and remains there as they disappear beyond the frame.

The relationship between the still and moving image is made explicit in Chris Marker's *La Jetée* (1962), which exploits the 'moving' nature of film to expose space and time structures. In *La Jetée* the film directs our attention toward the still image, the basis of the illusory cinematic 'moving' image and the primary construction of all photomedia. Used to extreme cinematic effect, *La Jetée* is made almost entirely of still photographs, with only a short scene showing moving images. These images are accompanied by a minimal soundtrack which consists of narration and a few sound effects. The narrator states at the start of the film that it is 'a story of a man marked by an image from childhood'.[59]

Set in the future, the protagonist travels back in time to save his civilisation. As he continually moves between his past and his future he meets a girl, sees his own death and finally dies at the same place the film begins, on the Jetée at Orly airport, the scene that marks the protagonist from childhood. These moments of his life and death are drawn together, each encapsulated by a collection of still images beautifully shot in evocative high contrast black and white. Individually, each image stands as a totemic signifier of a larger sequence of events which, in conjunction with the protagonist's telling of the story to animate the narrative, allows for the active participation of the viewer. Through this technique the film asks the viewer to consider the temporality of experience and the fragmented nature of photomedia and memory. Each fragment of the story told through a still image calls upon the viewer to animate it with their own understanding.[60] This is done by accessing one's memory of visual imagery, an inner-space that is connected to image-spaces. In *La Jetée* the still becomes the moving through the devices which animate the still. Marker's film is important because it deconstructs the materiality of the filmic medium into its core element: the still. It also reflects on our ability to access the past and project the future through photomedia. John Conomos claims that 'the human perception of time and space and its transformation through memory are the critical themes of *La Jetée*.'[61] Indeed this memory is part of the inner-space which connects with image-space.

In *La Jetée* light-time bears weight and functions like an image-world not unlike our own where one image can, through our memory of it, be understood as a totemic indicator of a larger narrative. *La Jetée* is another signifier of the emergence of light-space-time structures specific to the analogue era. Is it a function of the cinematic moving image and its accompanying 'still'.

Whereas *La Jetée* explores the still as the moving, in Andy Warhol's *Empire* (1964) the moving is explored as the still. The 8 hour, 5 minutes running time features a single continuous view of the Empire State Building in New York City.[62] Like *Wavelength*, the experience of optical reception is heightened and the physicality of vision becomes almost tangible as one's eyes search for some sign of change. Change does not come quickly; however, when it does come, it passes too rapidly to savour. Watching *Empire* is an experience of darkness and light, the statuesque Empire State Building emerges with luminous radiance from the dark night sky. Unwavering, the building is positioned firmly at the centre of frame. More prominent than the image, however, are the physical qualities of the film. The most pronounced 'action' in the film is the light flares which emerge occasionally, and the coming and going of watermarks that remain on the film stock from the development process it underwent. One's senses are heightened when viewing *Empire* as an awareness of the minutiae of time and the experience of watching become the dominant effects of the film. In *Wavelength* time and space seem almost tangible as one becomes highly attuned to the progressive nuances of the film and any changes and movement between one moment and the next. Whereas *La Jetée* lays bare the possibility of the still image to speak of events beyond those immediately represented within the image, in *Empire*, the moving *is* the still image in which the event is contained. This light-space modulation that occurs between the still and the moving image can be seen throughout the history of the moving image. From the first public movie screening on 22 March 1895, the moving image, like the still image, has formed and informed part of the global psyche.

Light-space in the analogue era

In the era of the early image machines photomedia was frequently used to bring micro and macro objects within the range of optical clarity. In this new era, the same application of technology continues; however, now it operates with immense speed. This additional function of photomedia and its light-space structures is discussed by Vilém Flusser and Paul Virilio, although their theories on this aspect are quite distinct.

While Flusser's theory of photography begins with the industrialisation of the image in the early era of the image machines, Virilio's theory of 'dromoscopy' begins in the analogue era, specifically in the late 1970s and early 1980s, as 'it was from that time onward that real time superseded real space'.[63] For Virilio, time becomes the dominant characteristic of visual experience, as approaching the speed of light visual information about an event occurs in 'the present', that is in real time. Whereas, previously, the time-lag between an event and news of an event was a function of the geophysical distance between the occurrence of the event and the recipient of news of the event, in the analogue era the event and its reception, via light, occur (almost) simultaneously. Geophysical space has become irrelevant. This is particularly so with television in which the event and knowledge of it is, to all intents and purposes, simultaneous.[64] He states that, 'currently, the systems and instruments of measure are less chronographic than dromographic, it is no longer the time of the passage that serves as the standard in a space passed through, but rather speed has become the privileged measure of both time and space.'[65] Virilio goes on to succinctly state that 'the speed of light does not merely transform the world it becomes the world.'[66] Again, Virilio highlights both the importance and dominance of the speed of transmission through light. Moreover, not just any light, but light which carries information of the photomedia image. The increasing application of photomedia directly represents the dominance of light in communicating the world and, more than this, in communicating new worlds through the image-spaces photomedia produces. It is this image-space, which forms through the fast transfer of images, that initiates the ubiquity of the photomedia image in the analogue era.

In contrast to Virilio, Flusser considers that the effect of speed is a bifurcation of space and time. Flusser contends that it is photography which causes a separation between time and space, claiming that 'these are the categories of photographic time and space. They are neither Newtonian nor Einsteinian, but they divide time and space into rather clearly separated areas.'[67] Flusser locates the 'photographic object', that is the object or scene the photographer captures, at the centre of space and time. Any potential view of the 'photographic object' such as a close-up or long distance view maintains its own time and space. The photographic act moves within the range of these times and spaces. Flusser claims that while it may seem that the photographer freely selects how they wish to take the photograph they are in fact only selecting one possible choice from the limited range of categories available. Therefore 'the freedom of the photograph remains a programmed freedom.'[68] Thus, Flusser would contend that all that remains in terms of photographic images and the image-spaces they form are images that show us the space and time categories that the photographer was engaged in when the image was taken.

You press the button: forming image-spaces everywhere

In 1888 George Eastman's company, Kodak, began manufacturing one of the most popular mass-produced cameras of the analogue era: the Box Brownie. Eastman's invention introduced the Kodak system in which a silver halide-in-gelatin dispersion was coated on a cellulose nitrate base that was loaded into a so-called 'fool-proof' camera.[69] The cost of the camera included the development of the 100 negatives it held and when all had been exposed, camera and film were returned to Rochester, New York, for processing. The packaging proclaimed, 'you press the button, we do the rest.'[70] Millions of people responded to Kodak's now famous claim by purchasing Box Brownies and in doing so increased the production of photographic images into the billions. Thus, a significant product of the industrialised world, the camera, proceeded to produce its own mass produced item, the photograph. With a large portion of the Western world involving themselves with the act of photographing potentially any moment in time, millions of lives were documented and catalogued within the family photo album and a plethora of images entered into a rapidly developing image-space. George Eastman had not only invented the Box Brownie, he had launched a new way of taking photographs: the snapshot, which as mentioned in Chapter 1, was a term originally coined by astronomer John Herschel in 1860 in reference to the essential nature of the camera to capture a small pocket of light-time. Now light-time was captured everywhere and the effect was felt as an image-space transmitted at an accelerating pace. The impact of the proliferation of the photographic image on society's perceptions and experiences was ostensibly subtle, but as history tells us, clearly profound.[71] A new relationship to the photographic image ensued. The physical world came to be known through the photograph and light-space evolved into a dominant structure.

In the analogue era society began to witness, 'a public which delighted in seeing the world pictured'.[72] Escalating industrialisation allowed for increased needs and uses for commodities; the photomedia image was available relatively cheaply to the masses and became one of these commodities. By the turn of the twentieth century photomedia was increasingly inserted into news media. Photomedia began to take on one of its contemporary manifestations: ubiquity. As Bernice Abbot states, 'our familiar awareness of the visual world has been evolved for us not alone by the eye but by the camera.'[73] Photomedia images, from the family album to the newspaper stand, were saturating society.

Flusser considered that the process of taking photographs for the purpose of representing our experiences meant that the photographic image acted to obscure the world. He believed that this occurred to such a degree that we fail to decode them and instead become intensely engaged and 'project them, still encoded, into the world out there'.[74] This aspect of Flusser's analysis implies that photography sits within society as an indicator of the formations of image-space and not necessarily the physical space photomedia is often assumed to represent. In short, for Flusser the production of the photograph produced image-worlds, or to use Flusser's terminology, a 'photographic universe' which mass culture accepted as an experiential reality.

At the movies: image-spaces in cinema

By the latter part of the twentieth century the still photograph was used to represent the effect of the moving image. Examples of this include the work of Jeff Wall who, like a film director, collaborates in the production of his still images, a process he calls 'cinematography';[75] and Ed Rusha who examines the filmic medium and the role of repetition in meaning, value and understanding. Two other separate but related examples of the effect of the moving image on perception can be found in Cindy Sherman's *Untitled Film Stills* series and Hiroshi Sugimoto's series of cinema interiors. In Sherman's *Film Stills* series, such as *Untitled Film Still* (fig. 2.6), the use of the female character and iconographic image production techniques particular to the classic cinema genre demonstrate a manifestation of image-space in the artist's psyche and public psyche. This photographic work and indeed Sherman's entire *Untitled Film Still* series portray scenes that, while not from any specific film, do come from the entire classic film genre as known to Sherman. In this way they come from the artist's inner-space and form a new image-space. The work is completed by the viewer who interprets and attempts to understand her images, they must access their own inner-space and form meaning through this. Influenced directly by Marker's *La Jetée*, the artist states 'I remember stumbling across a bizarre futuristic film that was made up of nothing but still images except for one of the final scenes which moved: Chris Marker's *La Jetée*.'[76]

Hiroshi Sugimoto's images of cinema theatres such as *Ohio Theatre, Ohio* (fig. 2.7) illustrate a phenomenological experience of watching films and its relationship to light-space-time. Sugimoto impresses light-space-time through his durational exposures of an entire film, which involves positioning his camera at the back of the

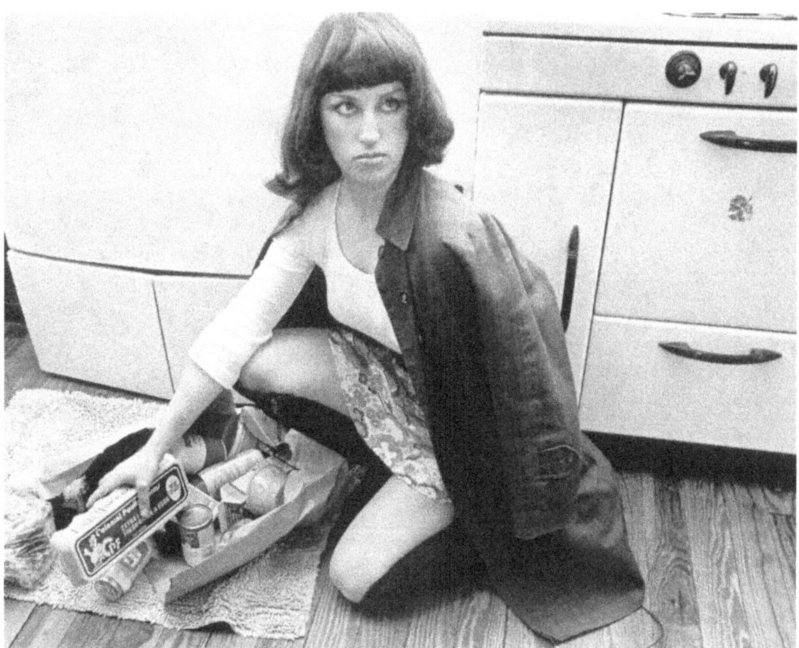

Figure 2.6 Cindy Sherman, *Untitled Film Still*, 1978.

movie theatre and leaving the shutter of his camera open for the duration of the film. Every still image, projected 24 times a second, merges with every other still image of the film so that in the final still image there is no figurative image of the film at all. What is captured by Sugimoto is the luminous glow of light compressed over time as it was emitted from the silver screen, and the sense of space that is created by this light reaching out into the movie theatre. Sugimoto recalls the first time he attempted this process.[77]

> One afternoon I walked into a cheap cinema in the East Village with a large-format camera. As soon as the movie started, I fixed the shutter at a wide-open aperture. When the movie finished two hours later, I clicked the shutter closed. That evening I developed the film, and my vision exploded before my eyes.[78]

Sugimoto's technique allowed the entirety of the film to become exposed onto the film frame, and although he would have understood the effect his long exposure would give him, the result,

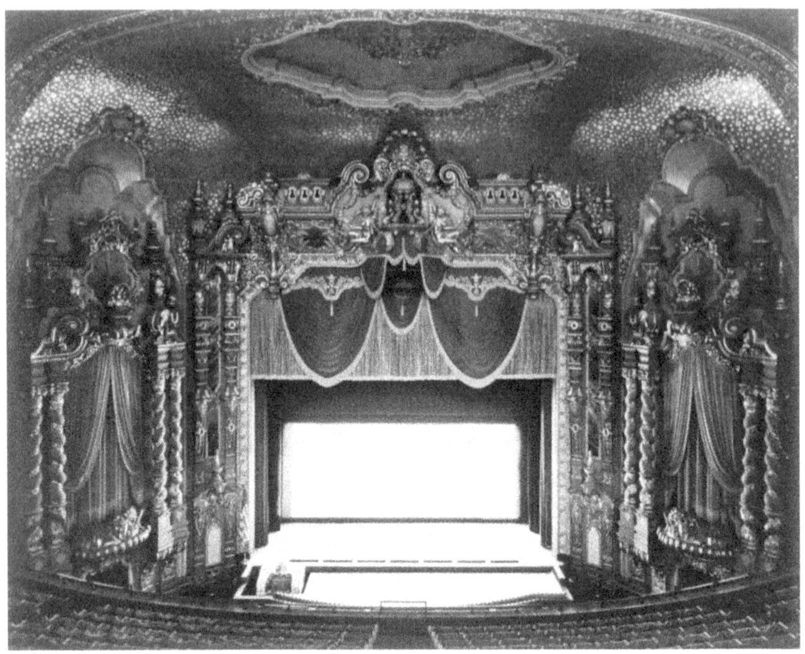

Figure 2.7 Hiroshi Sugimoto, *Ohio Theatre, Ohio*, 1980.

and what it conveyed, did not become clear until after he had developed the image. As David Campany highlights, 'Sugimoto's simple method enables us to think about film and photography as machines involving speed, light, exposure, projection, duration and motion.'[79] Sugimoto's process and the image produced embody these elements, allowing one to explore them fully through this image.

While Sherman demonstrates a direct expression of image-space, Sugimoto exquisitely illustrates the experience of this effect. Beyond these essential differences they both represent the ability of cinema and all photomedia to contribute to this image-space which cyclically moves between outer- and inner-space.

In fact, the proliferation of image-spaces does little to produce permanence through the image, despite the convincing ability of the still image to arrest light-time and convey light-space. Instead, as Flusser contests, this image-space is in a permanent state of flux where any image has the ability to displace others. Inner- and outer-space move in constant communication with a dynamic and evolving image-space. Flusser connects this re-use of photographic signs and signifiers to a magic ritual. To explain, he considers that photomedia, or 'technical images', do not signify the world but the program of the camera asserting that 'concepts no longer signify the world out there (as in the Cartesian model); instead, the universe signifies the program within the camera.'[80] In other words, the universe of images is programmed through the camera and as such signifies the technological program itself. Flusser ascribes each action as a 'magic ritual' of an eternally repeatable movement, claiming that 'technical images absorb the whole of history and form a collective memory going endlessly round in circles.'[81] This shifting image-space not only occurs through the snapshot or in cinema but through the entire gamut of photomedia; as Flusser states 'nothing can resist the force of this current of technical images.'[82] Flusser's understanding of photography and its relationship to memory is particularly apt when considered in relation to the documentary genre. He points out that 'there is a desire to be endlessly remembered and endlessly repeatable. All events are nowadays aimed at the television screen, the cinema screen, the photograph, in order to be translated into a state of things.'[83]

As Baudrillard contests in *The Art of Disappearance*,[84] the photographic process turns the world into objects – the world should not be considered as an already present object. On the contrary, as Flusser highlights, what little tangibility photomedia offers lies in its constant movement. These 'objects' are not fixed, their instability allows shifting and changing configurations of what is less like a stable object and more like a fragmented construct of realities: an image-world.

For Flusser two interweaving codes are present in any photograph. The first is that photographers, in an attempt at immortality, encode their concepts into the photographs they take, producing models for other photographers to use.[85] This is perhaps best demonstrated in the documentary photographs such as those taken by the cooperative Magnum[86] and will be explored shortly. The second is that the camera encodes and programs the photographic image in order to control society for the progressive improvement of the camera.[87] What Flusser means by this is that photography, as an apparatus, will continue to technologically improve because society is controlled by the apparatus to do so. This second code is without doubt the most radical of Flusser's claims and one that is difficult to dispute considering the continued and progressive improvement of the camera and the mass proliferation of the photographic image. Essentially Flusser's claim is this: better cameras are used to create more images to proliferate an image-world which both immortalises the photographer and continually advances photomedia.

Virilio gives a more foreboding forecast of the process of image propagation when he states, 'considered irrefutable proof of the existence of an objective world, the snapshot was, in fact, the bearer of its own future ruin.'[88] Virilio goes on to clarify this claim in simple terms, stating that, 'In multiplying "proofs" of reality, photography exhausted it.'[89] By this Virilio is suggesting that the space reality takes up is both consumable and finite and that the image displaces reality through its presence. Regardless, this does not stop an evolving image-space emerging from the analogue image machines. Some remarkable aspects of this era and its image-space can be found in so-called 'documentary' photography.

Proofs of reality

In 1947 photographers Henri Cartier-Bresson, Robert Capa, David 'Chim' Seymour, William Vandivert and George Rodger formed the cooperative Magnum. Initiated into photography and the associated image-spaces at a young age, founding member Cartier-Bresson in his book, *The Decisive Moment*, recalls taking snaps with his Box Brownie as a boy: 'from the great films I learned to look and see.'[90] Cartier-Bresson maintained a very different connection between the photograph and the world to that understood by Baudrillard, Flusser or Virilio. He expressed that 'to me, photography is the simultaneous recognition, in a fraction of a second, of the significance of an event as well as of a precise organization of forms that give that event its proper expression.'[91] Cartier-Bresson's claim that photography could capture, signify and express events is an important example of early twentieth century thought concerning documentary photography.

Vilém Flusser railed against the documentary style of photomedia believing that one who enjoys the structural complexities of the camera is doomed to be controlled by the apparatus. Those who are interested in continually shooting new scenes from the same well-worn perspective, such as those who take 'snaps' or practice documentary photography have not, according to Flusser, understood technical information. These photographers do not produce information or store moments of events as they intend. Instead, Flusser claims they 'prove the victory of the camera over the human being'.[92] This, he claims, occurs when the documentary photographer produces images that are what the photographer would deem technically proficient and unique but what Flusser sees as images, which demonstrate the program of the apparatus. Clearly, for Flusser this dichotomy between humans and machines is considered in terms of a power struggle in which the machine emerges victorious. However, Flusser forgets that the image machines of photomedia were in fact developed by human invention as part of a desire to control light. Through the capture of light, the images respond to and represent space and time characteristics and, in addition to their intended purpose, they change our understanding of light-space-time through the visual indication of space and time beyond our normal physical perception. The polarities of human and machine which Flusser critiques are better understood as an aspect of the human desire to contain and control light-space-time.

Documentary photography is, for Flusser, 'an image created and distributed by the photographic apparatus according to a program.'[93] His philosophy disregards the photographer as a free agent and states that the broad definition of a photograph is 'an image created and distributed automatically by programmed apparatuses in the course of a game necessarily based on chance, an image of a magic state of things whose symbols inform its receivers how to act in an improbable fashion.'[94] In other words, the photographic image created by the apparatus is a fortuitous arrangement of signifiers, which impel the photographer to make an 'image' from it. Cartier-Bresson, who by Flusser's definition is programmed by the apparatus and takes his photograph in the course of a game which depends on chance, is an exemplar of this scenario.

Cartier-Bresson famously claimed, 'there is nothing in this world that does not have a decisive moment.'[95] That is, Cartier-Bresson argues that a photographer hunts down, recognises and captures with their camera these 'chance' moments; that the photograph somehow stands for a larger sequence of events. According to Flusser this is not the case; the image only demonstrates the program of the apparatus and not a decisive moment.

Place de l'Europe, Paris (fig. 2.8) demonstrates Cartier-Bresson's 'decisive moment'. It both marks out a moment in time, and represents Cartier-Bresson's decisiveness in shooting that moment in time, transforming it into a photographic image. In this image Cartier-Bresson implies a larger narrative: what has gone before and what lies ahead for the leaping Parisian is not shown but can be extrapolated by the viewer through their own understanding of space, time and movement. Distinctly different to Marey's *Movements in pole vaulting* which explicitly identifies a moment in time through the physical path of the gymnast, *Place de l'Europe, Paris* goes further into light-space, connecting inner- and outer-spaces. Cartier-Bresson's image relies on one's understanding of the world to form the implied story for ourselves, as we animate it through past to future time. Inner, personal understanding connects with the image and a story unfolds through this connection. This image is contained but connects with time and space beyond the frame. Cartier-Bresson does not see this; he instead claims that 'this photography of the lens and shutter actively combines, colliding and colluding with the world in motion. The frame cuts into space and the shutter cuts into time, turning the photographic act into an event itself.'[96] Cartier-Bresson positions the act of photography as an event which cleaves space and time, turning it into an image; this image is a light-space-time structure which continues in space and time beyond the frame of the image and contributes to the shifting structures of image-worlds. *Place de l'Europe, Paris* is now a part of my image-world as it is a part of the worlds of others who see it. Many possible combinations and interpretations of these image-worlds may occur and through those multiple occurrences any singular notion of the real is destabilised. As Baudrillard states, 'photography also questions "pure reality".'[97]

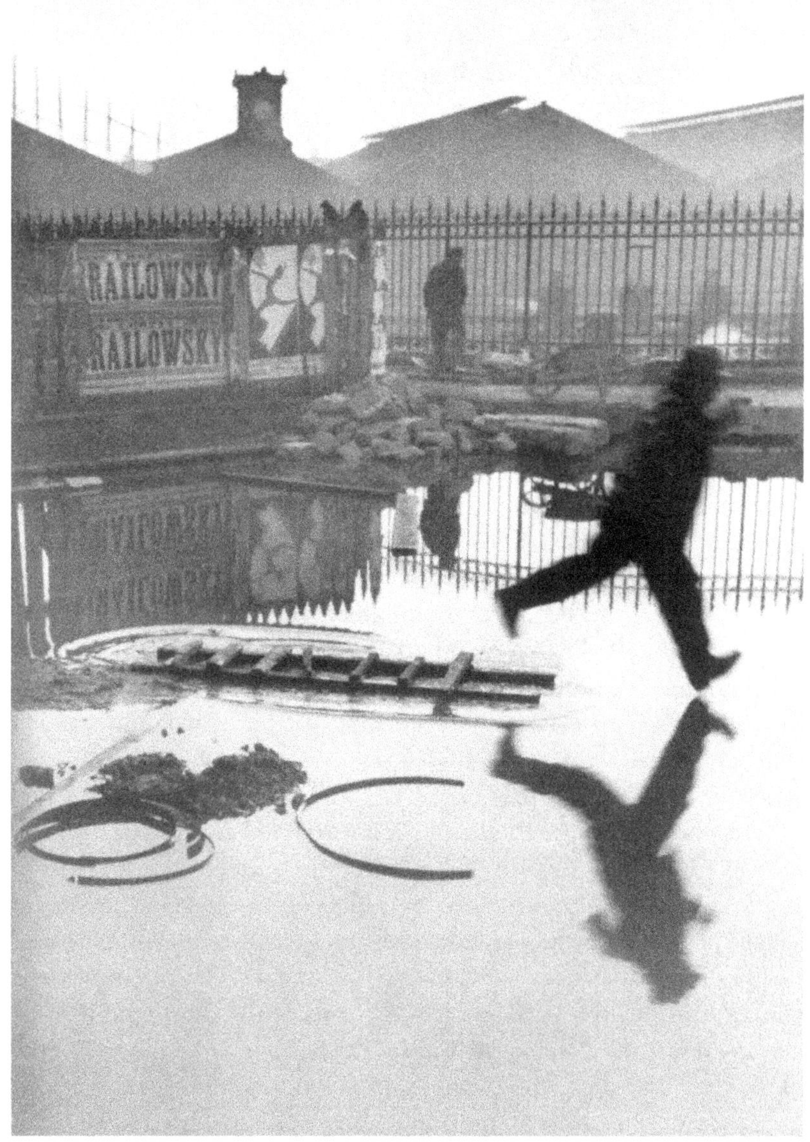

Figure 2.8 Henri Cartier-Bresson, *Place de l'Europe, Paris*, 1932.

Photomontage

The privileging of a singular moment in the history of photomedia is antithetically contrasted with the use of collage. The action of cutting became a useful strategy for artists when representing the emerging light-space-time structures of the era. As a technique, collage had been used as a craft for centuries before it was adopted by artists.[98] Sally O'Reilly suggests that artists took up collage as 'an attempt to address the prismatic Einsteinian universe that was emerging, where coherence and constancy were fast losing currency'.[99] Indeed collage artists also engaged with an increasingly fragmented society, particularly during and after World War I.

Pablo Picasso and Georges Braque were precursors of the process as a strategy for art making in the early 1900s. By the 1920s photomedia was incorporated into collage as photomontage[100] by artists such as the Berlin based Dadaists John Heartfield, Raoul Hausmann, Hannah Höch and George Grosz. Implemented for a variety of reasons such as formalist composition or for political critique, collage was associated with various art movements including Cubism, Dada and Surrealism and later used to address issues ranging from feminism to mass media critiques. Michael Frizot states, 'the unifying principles of photomontage developed in a variety of different ways according to the artistic movements that adopted it, in view of the possibilities for reform, protest or style that it offered to all.'[101]

In photomontage the image is layered and fragmented with a multiplicitous simultaneity of initially disconnected and at times random images culled from the daily flow of the burgeoning image-spaces produced by mass media outputs such as magazines, books, newspapers, advertising and catalogue images. All of these were contributing to the light-space structures of the analogue era. As David Lillington claims, 'the found image starts with the "twilight" of an image and can be thought of as its double: metaphorically, its shadow.'[102] Lillington is suggesting that by resurrecting the image at the end of its initial circulation and life it gains a new life separate from but still connected to its origin. This fracturing of the image creates a further temporal fragmentation: past use compared to current use. In the first half of the twentieth century John Heartfield was producing work which mirrored the disdain and outrage at the rise of Hitler's Third Reich. *Hurrah, die Butter ist alle!* (1935) which translates to *Hurrah, the butter is all!* is a prime example of photomontage. Below the title is a quote from Hermann Göring made during the German food shortage. Translated, the quote reads '*Iron has always made a nation strong, butter and lard have only made the people fat.*' This image with its clear political critique of Nazi Germany is indicative of early photomontage. The images were sourced from newspapers and recombined for their potential points of connection and to parody the aesthetics of Nazi propaganda images. In this image a family sits down to a meal of industrialised metal, even the dog enjoys a large bolt while the youngest member of the family feasts on a machine gun. A picture of Hitler hangs in the dining room, the walls decorated with swastika motif wallpaper.

The photomontage technique like that of collage involves a simple action of cutting and reconfiguring the image with other excised images. This action of cutting, which may draw an obvious comparison to trauma, can also be connected to the symbolic act of intervention or resistance, where the self-directed narrative is positioned as superior to a passive acceptance of the supplied one. Whatever the case, photomontage is used for a variety of reasons and does not necessarily indicate trauma, revolt or any other dramatic scenario. It may simply be an alternative created for its own sake; yet another example of participation in a ceaselessly forming image-space where every individual can, potentially, contribute their own perspective. Perhaps closer to the idea that the image does not directly represent the world, the photomontage process of cutting and rearranging the photomedia image in unexpected ways is akin to the actual experience of the increasingly fragmented nature of space and time. As Sally O'Reilly explains, 'the induction of collage into art coincided with enormous cultural and epistemological shifts in which speed, acceleration and flux factored greatly.'[103]

A highly significant technological contributor to the speed and flux of the analogue era is the cathode ray television. Initially invented at the turn of the twentieth century, by the mid to late

1950s it was the dominant form of home entertainment. The live telecast was a revolutionary step forward for photomedia technology. Events were carried at the speed of light into the domestic space. Television was extended further with the ensuing invention of portable video equipment. The video camera quickly came to have more general applications than its initial purpose as a surveillance aid to the US Army during the Vietnam War. As Virilio claims, 'the speed of light becomes the absolute weapon and the light of speed produces the perfect image, the hologram of pure power.'[104] Soon the video camera was not only used in war but in the home and everywhere else too, further contributing to the proliferation of image-spaces.

Collage gained new relevance with the introduction of television and what is dubbed 'the first living room war': the Vietnam War. This televised battle provided a new development in the relationship between inner- and outer-space though image-space. Artists again took up photomontage as a strategy for exploring these image-worlds. Just as Heartfield and the other early photomontage artists used images that were a part of the existing image-space so did artists such as Martha Rosler 30 years later. In *Red Stripe Kitchen* (fig. 2.9) Rosler combines fragmented spaces of 'there' (the war zone in Vietnam) and 'here' (the domestic interior of a kitchen). Her use of documentary images critiques the increasing reliance on documentary-style images to inform, or to represent what was happening in Vietnam. As Alexander Alberro claims, Rosler's work features 'an engagement with traditions and practices of documentary and their ability to impart information, as well as the sorts of information imparted'.[105] Rosler uses the fragments of information she collects and reconfigures this information to represent a new message, her image-world. Unlike Bresson who sought to 'cut' the image from the narrative surrounding its origin, Rosler's 'cut' is a juxtaposition of an illogical combination of images to create an unpredictable and challenging narrative structure. Ian Munroe discusses the 'cut' in terms of the 'edge' as an active boundary, a contained location in which previously disparate images cohere.[106] Dislocation, relocation and transformation are typical acts of the photomontage artist in the formation of colliding narratives and improbable relationships.

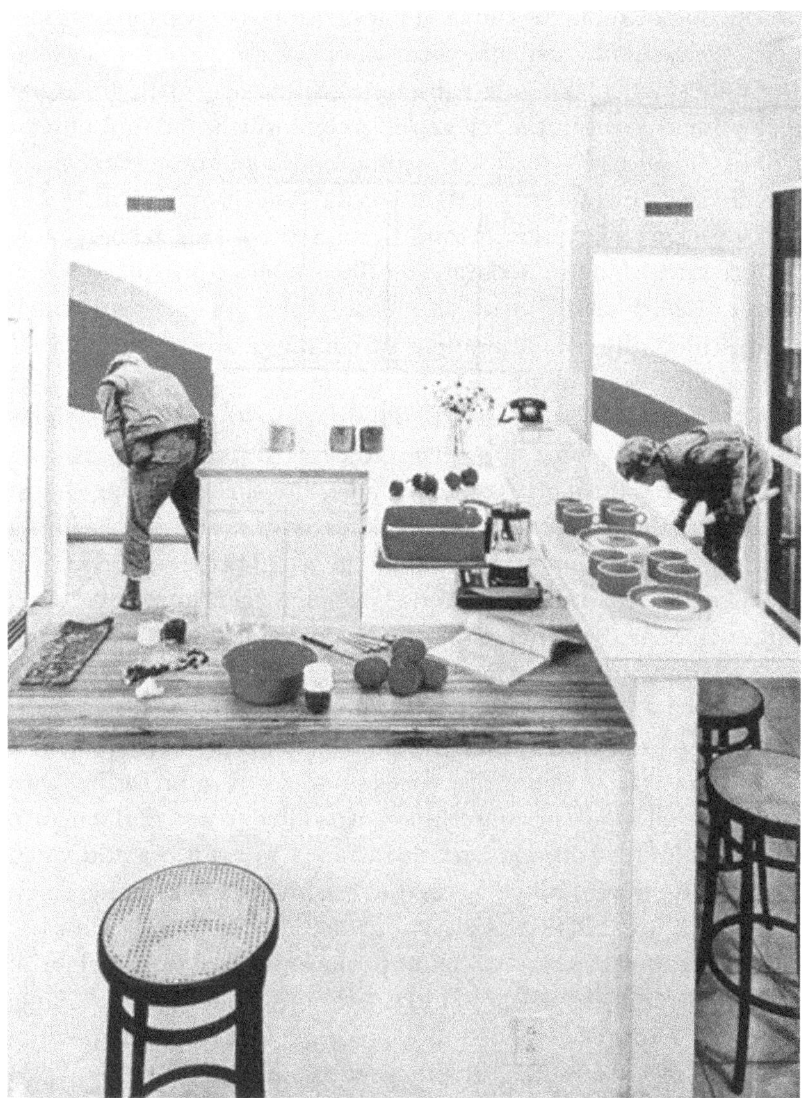

Figure 2.9 Martha Rosler, *Red Stripe Kitchen*, 1967–72.

Video Art

In the same way that the invention of television motivated artists, so did the invention of the video camera. With television and video, light is emitted from and not onto the surface as it is in the production of analogue film and photography (see Chapter 3). The first commercially available video cameras were sold for between US$1,000 and $3,000 – a substantial difference to their closest technological cousin the television camera which sold for between US$10,000 and $20,000.[107] The most popular and prominent video camera to come onto the market was the Sony Portapak in 1967. A bulky yet portable video recorder powered by large rechargeable batteries which were separate to the camera and linked via a power cable, it carried black and white, reel to reel magnetic video tapes which allowed 20 minutes of continuous recording. Other less popular yet similar cameras also available at the time were the Panasonic Concord and the Philips Norelco.[108] These systems were small enough for one person to carry and control, providing a new liberty through their portability. At this early stage most amateurs edited 'in camera'[109] as access to a professional editing studio was not common. Soon a new genre of art emerged known as video art. Artists such as Andy Warhol,[110] Nam June Paik[111] and Buky Schwartz,[112] among many others, responded quickly. With the new medium came the introduction of a new way of working; it provided artists with new ways to express and explore the physical body in space, which was at the time a politically and artistically important topic.[113] Politically, issues of power were prevailing with women's rights and the conscription of 'bodies' to war. Performance art, happenings and body art also emerged in this era and video along with photography was used to document some of these new ephemeral art practices. Early video art also included themes of space, time and materiality, seeking to move away from representative techniques and paradigms found in previous models of art making. Video art was a new opportunity to engage and extend issues and concerns specific to the period in which it emerged.

The Sony Portapak, which became the camera of choice for artists, was first used by Nam June Paik. (widely acknowledged as a seminal artist in defining the medium in 1965.)[114] He engaged with video and the screen as a metaphysical experience, where the phenomenological relationship with the video object (screen and image) in time is the heightened experience of the work. Paik situated the medium between art object and performance. His *Moon is the Oldest TV*, 1965, is a keen observation of the power of televised media, through its reductive image format.[115] The installation consisted of 12 television screens mounted on black plinths and sequentially positioned to show an image of the Moon, in various phases of the lunar cycle as seen from Earth. As with Whipple's *Moon* (1851), Paik reveals the fundamental relationship between photomedia and light-space and its capacity to traverse geophysical space and bring 'there' closer to 'here'. Light-time is represented through the successive phases of the Moon, both as static images and through the unfolding of time indicated by each Moon's phase. However, these images of the Moon are in fact circles and crescents of light; each cathode ray tube in each TV has been modified by Paik to produce this effect. The representation and belief that the image is of the Moon is constructed by the viewer, if the viewer understands each image to be a Moon then that is what it is. If the viewer understands each image to be a circle of light, then that is what it is: such is the illusory quality of photomedia. 'Paik's reductive reconstruction of the television becomes an element in a conceptual installation of the medium and rendering of its defining qualities: temporality and the changing image.'[116]

Another artist who used early video technology was Andy Warhol. As a prototype the Norelco video camera was delivered to Warhol's Factory for his experimentation. Warhol used it to shoot part of his video installation *Outer and inner space* (1966) featuring Edie Sedgwick. As with *Empire*, Warhol uses the raw materiality of his medium, this time both film and video, to explore time and space. The sense of immediacy, which is a central aspect of video, is understood by Warhol and explored in this particular work. The process of film and filming is highlighted through Warhol's doubling up of Sedgwick. Initially filmed once, Sedgwick is later placed in front of the new Norelco video camera and filmed again; this time while the sound of her voice and image from the original film is played back to her. Sedgwick is not an isolated performer alone with the camera, she interacts with people off camera responding to cues, and at one time sneezing in time with her filmed image. In these moments the filmed outer space of the recorded light-space-time and the inner space of the unseen psychology of Sedgwick are brought together. Sedgwick is positioned so as to encounter these two aspects of herself, outer and inner, at once.

Rosalind Krauss wrote about early video art, drawing a comparison between the role of mirrors and video. Krauss wrote that mirror reflection 'implies the vanquishing of separateness. Its inherent movement is toward fusion. The self and its reflected image are of course literally separate. But the agency of reflection is a mode of appropriation, of illusionistically erasing the difference between self and object.'[117] In *Outer and inner space* Warhol constructs an image-space in which Sedgwick is left to examine her presence as simultaneous object and subject.

Buky Schwartz's *Yellow Triangle* (fig. 2.10) is an excellent demonstration of perceptual space and the relationship between the viewer and their understanding of this space. In an extension of the analogue era's light-space structures Schwartz creates an image-space which is equally physical and immaterial.[118] In *Yellow Triangle* parts of a gallery are painted in such a way that from one particular vantage point a yellow triangle is visible. A closed-circuit video camera is positioned at this vantage point and two monitors placed in the space show video footage of the room. The yellow

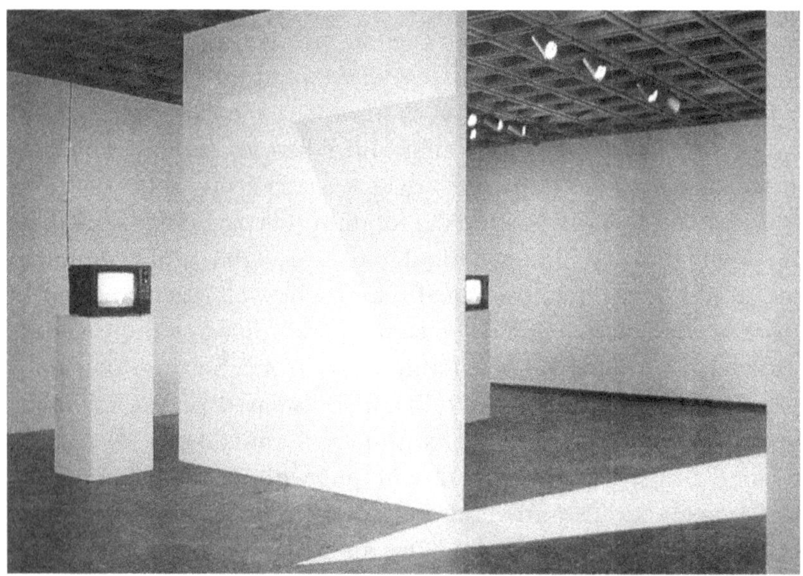

Figure 2.10 Buky Schwartz, *Yellow Triangle*, 1979.

triangle, which can be easily seen on each monitor, transforms the depth of the gallery space into a two-dimensional pictorial space. On entering the gallery the viewer has the opportunity to be engulfed by the yellow triangle and simultaneously become part of the picture. This allows multiple perspectives to combine, that of the viewer and that of the video camera. As John Hanhardt writes, 'the spectator encounters *Yellow Triangle* on two levels, in the physical space of the actual painted triangle, and through the electronic point of view of the video camera/monitor screen on which the triangle is seen.'[119] In this work and in many of his other 'videoconstructions'[120] Schwartz collapses the distinction between physical and non-physical space by placing the viewer within the physical gallery space and at the same time presenting another space only perceptible through the video monitor. *Yellow Triangle* is an illusory space where fragmented parts of space and the geometric shapes such as that of the triangle and the gallery connect through the configuration of perspectives and positions. This is similar to the photomedia initiated image-spaces, where specific configurations and perspectives create realities and understandings.

Virilio remarks that 'Since every object is for us merely the sum of the qualities we attribute to it, the sum of information we derive from it at any given moment, the objective world could only exist as what we represent it to be and as more or less an enduring mental construct.'[121] Anne-Sargent Wooster proposes that Schwartz's 'video constructions' are not only a hybridised art form, new to the era, but that they 'offer a valuable harbinger of future realities'.[122] Indeed, Schwartz's *Yellow Triangle* is not only illustrative of the analogue structures of light-space but a prophetic representation of the increasingly fragmented nature of image-space that will be manifested in the digital era (discussed in Chapter 3). Schwartz's use of a geometric form is pertinent as it gives this image-space a structure which is individually determined by the artist. Virilio claims cinema relates to architecture, stating that 'Cinema provides *matter* for simultaneous collective reception the way architecture has always done.'[123] Many people can receive the same procession of information at the same time through architecture; it is no surprise that the reception of photomedia is at times structured and navigated in a similar way. Schwartz's work is both cinematic and architectural but, what happens when the physical structure, the tangible aspect of the image-space and the outer-space, reforms, as it does constantly, and retreats from its material presence towards the immaterial?

Leap into the void

Yves Klein's *Leap into the void* (fig. 2.11) represents the transition from one material space into another; it is also a representation of Klein's inner-space transformed through photomontage and made manifest in outer-space.[124] For Klein this was a utopian space where human flight was possible. The image was created at a time when the first space vehicles were launched. In *Leap into the void,* Klein is freed from the gravitational laws of Earth and is suspended in transition leading towards the void. Through a combination of elements from Occidental and Oriental philosophy Klein constructed his model of the void. From Western philosophy Klein used the concept of transcendence from one space into a higher realm, much like the Christian model of heaven,[125] and from Eastern philosophy, concepts of nothingness and of loss of individuality in favour of an eternal existence in touch with everything. In *Leap into the void* this is demonstrated as Klein moves away from earth and into the sky. What is at play in Klein's void is space as immaterial nothingness. Through the void and its implied nothingness the potential of the future and its absence in the now, as well as an absence of certainty, is implied. Beyond *Leap into the void,* Klein also expressed his concept of the void through his monochrome paintings, sculptures and performances which 'offered an unmediated experience with nature, with the universe, and, paradoxically, with raw matter'.[126] Klein aimed to connect with the matter of being. The essential nature of experience and his art purposefully sought to either demonstrate or invoke this in the viewer. In *Leap into the void* Klein is represented as making his way into the void while the cyclist below continues along his set path, unaware of the possibilities that Klein is moving towards.

Paul Virilio makes similar links between the progressive developments of society and the loss of physical space when discussing his theory of 'dromology'. Virilio proposes that through the action of propulsion and the excess of speed that results, the passenger (the viewer of photomedia) may cross over the void into

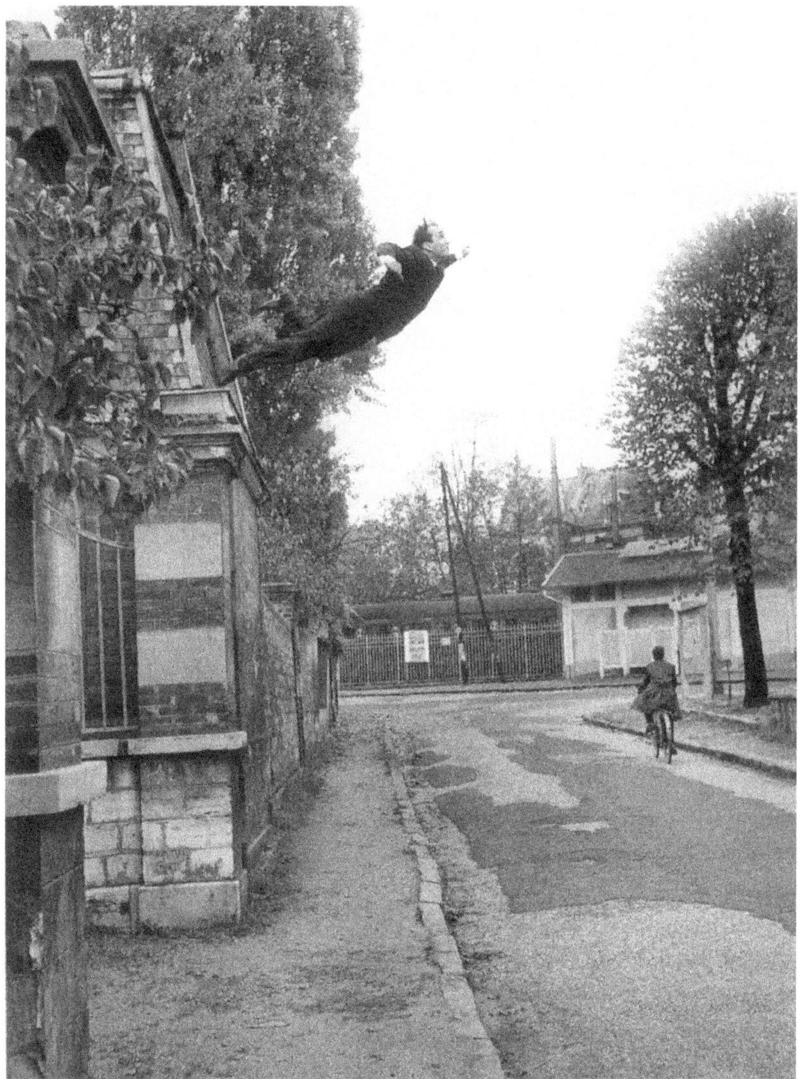

Figure 2.11 Yves Klein, *Leap into the void,* 1960.

the atmospheric volume above the surface of the Earth.[127] By this Virilio is referring to the act of transcendence of the physical world that may occur though psychological connections. Relating the effects of 'dromology' to motor travel, Virilio cautions that such an action results in a loss of duration and experience; only departure and arrival maintain their integrity. In the future Virilio predicts we will be reduced to the experience of arrival only as 'each of us remains in our own place while awaiting the arrival of the transmission (telephonic or televised) ... Dromoscopy is, therefore, paradoxically *the wait for the coming of what abides.*'[128]

For Flusser a void forms in the transition from the analogue era to the digital era[129] and from where the photograph remained as a post-industrial object tenuously connected to the thing it represented. He claims that 'as long as the photograph is not yet electromagnetic, it remains the first of all post-industrial objects. Even though the last vestiges of materiality are attached to photographs, their value does not lie in the thing but in the information on their surface.'[130] In this state objects are replaced by information, while Baudrillard, whose philosophy is intensely tied to the idea of a void, considers that, 'preoccupied with creating a void, we will not tolerate much longer the density of matter.'[131] This is the crux of the issue: for each of these theorists a loss of physicality or materiality translates into the construction of a void. This void however is not empty as it might seem but full of many possible realities, like an image-space.

Virilio, Baudrillard and Flusser all consider the future to be a void, an idea which gains support in the digital era and will be discussed in Chapter 3. So, in this transitional space, at the last moments of the analogue era and at the start of the digital era, what are the essential changes?

Progressing from this analogue era, Flusser claims photomedia contains a new 'magic'.[132] This moves us closer to a world in which reality retreats and is lost in an image world, which for Flusser is somewhat like a void in that it has no connection to physical matter. This idea forms part of Flusser's 'spell' theory whereby we are, to all intents and purposes, under the spell of the program – the photomedia program – and while under this spell we will work in the interest of perpetually improving the photomedia machine.

Virilio connects the proliferation of the photomedia image to an increase in speed. If speed is light, all the light of the world, then what is visible derives both from what absorbs and reflects light and the appearances of momentary transparencies and illusions. The dimensions of space are themselves only fleeting apparitions, in the same way that things are visible in the instant of the trajectory of the gaze: this gaze that is both the eye and defines place.[133] In other words personal perception creates the reality. 'The only unit of measure is therefore absolute speed, produced by the absolute temperature: the speed of light.'[134] At this speed we reach the luminous void, a light-space.

In spite of the rapid and profound changes that occurred in photomedia, the man who had been alive to see much of this progression, Henri Cartier-Bresson, offers one insightful comment. He saw any changes to photomedia as inconsequential, stating that 'photography has not changed since its origin except in its technical aspects which for me are not a major concern.'[135] More to the point Baudrillard claims that 'no matter which photographic technique is used, there is always one thing, and one thing only, that remains: the light.'[136] And the intrinsic light-space-time structures.

3

DIGITAL IMAGE MACHINES C.1990–2013

Dream machines: technology at the speed of light

Since the era of the earliest image machines, photomedia has progressively developed more mechanisms to technologise light in service of the production of images. The digital era sees the replacement of analogue storage systems such as film and magnetic tape with digital storage systems such as the charge-coupled device (CCD) and the active pixel sensor (APS). In analogue image machines light is stored on film coated in an emulsion layer containing grains of chemicals. The film is loaded into the camera and requires chemical development, processing and printing to render viewable the photographic image. In digital photography the storage device is an image sensor which transforms the received light into digital information; this information is then processed as an array of pixels which denote specific colours and hues made visible with the aid of a screen to render viewable the photographic image. In both forms light is stored using a physical transformation of a layer on a substrate ready to be made visible once again through development chemicals or digital techniques before the use of appropriate viewing mechanisms such as a print or the screen. When considered at that reductive level, the distinction made between analogue and digital (chemical grains or pixels) becomes little more than a matter of aesthetics.

The shift from analogue to digital has been rapid,[1] even taking the largest commercial producers of photomedia technology by surprise. Many of the primary producers of photomedia technology that dominated the market in the twentieth century such as Kodak, Nikon and Polaroid suffered financial losses at the start of the twenty-first century because they did not accurately predict the speed of change toward digital technology. Polaroid provides a case in point: the company was founded in the 1930s and began production of their most popular and well-known product, the instant camera, in 1948. By 1991, the same year that the first digital camera was released onto the commercial market, the Polaroid instant camera was at its peak popularity, earning the company US$3 billion in revenue. Within ten years digital technologies had come to dominate the market and Polaroid, failing to embrace this change, declared bankruptcy.[2] However, the Polaroid brand continues today largely because of nostalgia for their original products and less so because of any advances they make in the digital technology.

The genesis of commercially available digital photomedia began in 1991. That year saw the release of the Kodak DCS-100,[3] a camera aimed at the photojournalism market. Although primitive compared to current basic consumer cameras which on average shoot images with a resolution of 16 megapixels or more, the DCS-100 achieved Kodak's primary goal to accelerate the transmission of pictures back to the studio or newsroom.[4]

Today, the capture of images by consumer photographers can achieve a standard resolution of between 10 and 21 megapixels for most cameras. At the same time digital video capture has been enabled by small, portable and sophisticated cameras. Currently, storage is supplied in terabytes, equating to the possibility of many thousands of still images and many hours of video captured on a single storage device.[5] It is already apparent that the digital era provides us with an unprecedented capacity for making and seeing images.

Image-spaces of the digital era

'All that we see or seem'[6]

Since the early image machines, photomedia has been used to create images which manipulate and construct realities.[7] Examples of this can be seen in the tableau of Hippolyte Bayard's photograph *Self Portrait as a Drowned Man* (1840), the constructed image of Henry Peach Robinson's *Fading Away* (1858), which used multiple images combined to form the final image, and Georges Méliès's 1902 classic film *A Trip to the Moon*. Staging an image or constructing one from multiple, separate images is not at all new. For Robinson the only methods available for manipulation of the image were comparatively crude and limited to the physical parameters of cutting into the negative to alter composition, or exposure techniques such as dodging and burning. Now, an artist working with digital photomedia expands these basic structures of image manipulation with ease and speed. The primary technological facilitator of this is the conversion of light into digital data. As digital data the image can be easily transferred to a computer which allows the transformation of this data and in turn the image. The scope of control of digital data is tied to computer technology which is rapidly improving, and thus constantly increasing possibilities for image construction.

For instance, Jeff Wall's photographic practice began in the late 1970s well before digital photomedia came about. His body of work can be divided according to his two distinct approaches to his 'cinematography'; the first, 'documentary', involves Wall photographing found scenarios and objects; the second, 'tableaux' uses sets and staged scenarios. Both approaches may involve many days, weeks or months of shooting the one scene; however, the final process involves carefully constructing the final image through the composition of many images, a method similar to that of Henry Peach Robinson.[8] These images are created from memories of Wall's daily encounters or from formal influences such as cinema, painting, theory and at times, literature. Since 1991, Wall has used digital technology to facilitate his practice of combining many separate images – at times hundreds if necessary to support his painstaking compositional approach. Wall's use of multiple images re-combined into the single, final print is much like the technique used by Robinson. The idea was first considered by Hippolyte Bayard when proposing a solution for adequately exposing a landscape and sky scene. However, it is likely that Wall's idea for this technique arose in the early 1970s after research for his unfinished thesis on the collage artist John Heartfield. Wall's images are highly detailed, large-scale transparencies displayed as a lightbox.[9] This presentation format endows the image with a luminosity and presence within the gallery space.[10] The image, because it is a transparency, channels light from the light source inside the lightbox, giving the impression that the image is glowing. Light tangibly moves through the image before emanating into the gallery space.

In *After 'Invisible Man' by Ralph Ellison, the Preface*, 1999–2000 (fig. 3.1) Wall takes Ralph Ellison's 1952 novel *Invisible Man*[11] as a narrative for his photograph. *Invisible Man* is Wall's third image based on a literary text.[12] In it we see the protagonist sitting on a chair, and while his face is not visible his posture is suggestive of a contemplative state of mind as he bends forward, drying a silvery pan. The room he is in is cluttered with items and in the centre of this room rests a lounge chair, record players, clothing, rugs, bowls, cups; myriad objects this man might need to comfortably live within this room. All of his many belongings appear to be from the 1950s, the same decade that Ellison's story was penned. As in Ellison's story, the ceiling is covered with 1,369 illegally connected light bulbs. Ellison's character declares, 'Perhaps you think it strange that an invisible man should need light, desire light, love

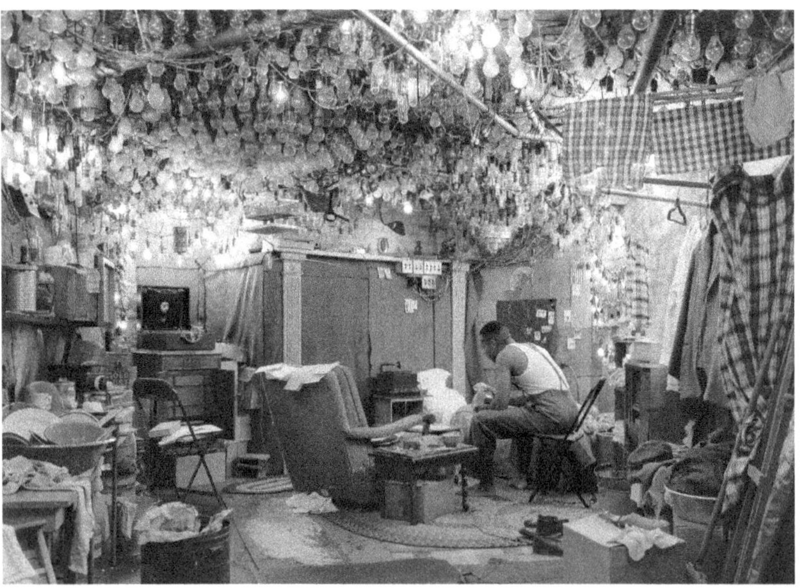

Figure 3.1 Jeff Wall, *After 'Invisible Man' by Ralph Ellison, the Preface*, 1999–2000.

light. But maybe it is exactly because I *am* invisible. Light confirms my reality, gives birth to my form…Without light I am not only invisible, but formless as well; and to be unaware of one's own form is to live a death.'[13]

There is then a thematic corollary between Ellison's story, which uses light as a way to discuss race relations and an associated visibility within 1950s America, and Wall's practice, which uses light as an essential property of 'seeing' this image – from the use of light in all aspects of creating the photograph to the display of this work within a lightbox. Wall's use of light in the construction of his images is part of his quest for realism, an attempt to transform his 'vision', to bring his imagined inner-space, into an image, an outer image-space. Digital photography assists Wall in obtaining realism. As he argues, the sharpness of the final composite image is closer to what the eye perceives.[14] Having often cited the importance of Charles Baudelaire's theories in regard to his own work,[15] in *Invisible Man,* Wall attempts to emulate Baudelaire's view of realism as outlined in *The Painter of Modern Life* (1863).[16] Baudelaire claims 'The Painter of Modern Life' extracts moments from his normal experiences of the world and distils them into his work. As Michael Friend explains, Wall does not present a fantasy world of his own making; *Invisible Man* is a representation of, 'a shared world, inflected individually'.[17] Wall presents his personal inner-space view of an outer image-space world which we all inhabit. In this sense *Invisible Man* not only represents the world Wall has observed but also represents the shared image-spaces of the digital era.

Like Wall, Gregory Crewdson photographs highly staged and cinematically lit scenes such as *Untitled 'Beneath the Roses'* (fig. 3.2). Also like Wall, Crewdson's final composited image is created from many images that are further refined and directed in a post-production digital environment. But, while Wall actively works to represent reality, Crewdson's dreamlike photographs reference the Freudian underworld of the human psyche by presenting a series of signifiers for the viewer to bring their own associations to.

In *Untitled 'Beneath the Roses'* the viewer is presented with an almost empty, nondescript street of a small American town, possibly North Adams or Pittsfield, Massachusetts, where Crewdson shot his series *Beneath the Roses.* These archetypal towns, built on industry,

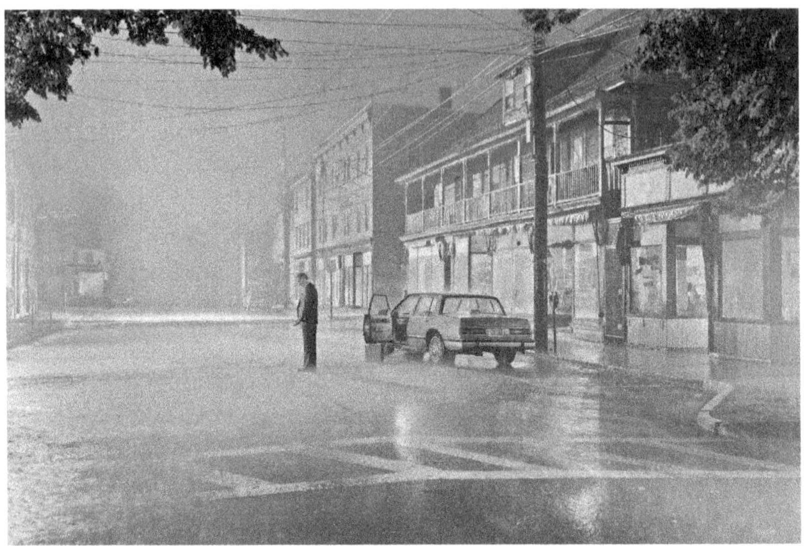

Figure 3.2 Gregory Crewdson, *Untitled 'Beneath the Roses'*, 2004.

were once beacons of industrial progress, but now they are slowly decaying. These towns form a 'set' for Crewdson's series which depicts a community that is 'corrosively lonely'.[18]

In *Untitled 'Beneath the Roses'* a car has stopped, the driver's door is wide open, a briefcase sits beside it, on the road. A man in a dark suit stands in the rain, looking at his hand. It is difficult to see his expression, but his gesture suggests he is lost in thought, much like the protagonist in Wall's *Invisible Man.* In *Untitled 'Beneath the Roses'* it is not clear what has occurred; Crewdson does not reveal such details. This image does not clearly relate to a familiar narrative, instead the moment 'hangs' separate from any before-and-after narrative arc; the image is suggestive of an indeterminate 'story' beyond the frame. The combination of referents within a single, still photograph allows the viewer to bring their own 'story' to the image; memory can mingle with perception, facts easily blend with fiction; a scenario is suggested but the larger narrative is left open. The use of a familiar language to connote a sense of memory aids multiple readings. The power of *Untitled 'Beneath the Roses'* is procured from Crewdson's use of symbols and gestures derived from American cinema. They are from no particular film but, due to the mystery they evoke, have a sense of the film stills of Spielberg, Lynch and Hitchcock.[19] Crewdson's images probe a subconscious 'bank' of images, or image worlds, which American cinema has impressed on the collective psyche.

Untitled 'Beneath the Roses' is a strange but powerful mix of the artificial, the prosaic, the dreamlike and the real, combined with a sense of nostalgia. The image is not only made so that the viewer might bring their own reading to it, but so that Crewdson himself can 'see' the image. Russell Banks considers Crewdson's intention, in his laborious process of creating images, to be 'not so that we, his audience, can see what would otherwise be invisible to us, but so that he himself can see'.[20] In a sense Banks is correct: Crewdson's images are an act of visualising the personal inner aspect of image-space. However, if these images were purely for Crewdson's own benefit he would perhaps not show them to others, yet he does, and in doing so contributes to the cyclical nature of photomedia image-space. Image traces in public circulation are exploited by Crewdson to realise his inner image-space. In turn, as he projects it onto the public image-space the image is then absorbed into the subjective image-space of the viewer in an endless replay of images. In a sense, what we are witnessing is the development of a 'language'. As the lexicon of image-signifiers mutates and expands so too the possibilities of image-spaces increase and, much like a library, more image-spaces are constructed, read and circulated throughout the visually literate reading public. The movies and movie stills that Crewdson and much of Western culture have long experienced form part of his understanding, which he then projects as new photomedia images that form part of the always shifting image-space, in a cyclical movement between private and public. As Crewdson points out, 'an artist's pictures also carry their own meaning once they're out in the world.'[21] This work is part of the image-space that has been proliferating since the invention of photography.

Wall's and Crewdson's images demonstrate the way light-space is received, physically by the body, to inform perception and form an inner image-space that is then projected 'out there' through photomedia images. This Flusserian cycle is a process of perpetually receiving and in turn contributing to the trajectory of this image-space and so creating a 'photographic universe'. The new images received 'in here' come to form new images, 'out there'. Just as Wall's *Invisible Man* represents 'a shared world, inflected individually',[22] so does Crewdson's *Untitled 'Beneath the Roses'* and so do all photomedia images. With the speed and flexibility the digital era now provides, these image-spaces have gained prominence through their creation and dissemination as digital information.

While Wall and Crewdson use digital technology to extend their image production techniques, many artists use the new capabilities to not so much realise an already existing reality but to make real a fantasy that is not yet a reality. Patricia Piccinini, Erwin Olaf, Sam Taylor-Wood and many more have all used digital technology to produce images of their versions of a 'possible' or 'phantasmagorical' reality. These artists use digital photomedia to construct images that are in parts prosaic, spectacular and specular. They touch on an important aspect of the digital era – to challenge the long held notion that for something to be real it must exist in real space. Instead, they propose that for something to be real it only needs to be understood as such. And, since the photographic image tends to be viewed as an objective, veridical representation of 'reality', so their 'fantastical' constructs are taken to be real.

Charlie White's image *Champion* (fig. 3.3) provides a case in point. Here a young man stands in the middle of a room holding a giant head dripping with blood. These characters look directly at the camera and thus acknowledge it, and perhaps, their place within the malleable composition of the image. Their gaze also serves to acknowledge the viewer as if they too are there in this space; this helps to position the viewer within the image. In *Champion* White brings together the real and the imaginary. In this instance it is the fictive biblical figures David and Goliath (the image a reference to Andrea del Verrocchio's bronze statue, *David*, 1473–1474), that are composited into a real room and space. Like Verrocchio, White has placed his own facial features on both heads, David a representation of the artist as a young man and Goliath as the artist much later in life. The fragments taken from the physical world are recombined to form this fantastical image.

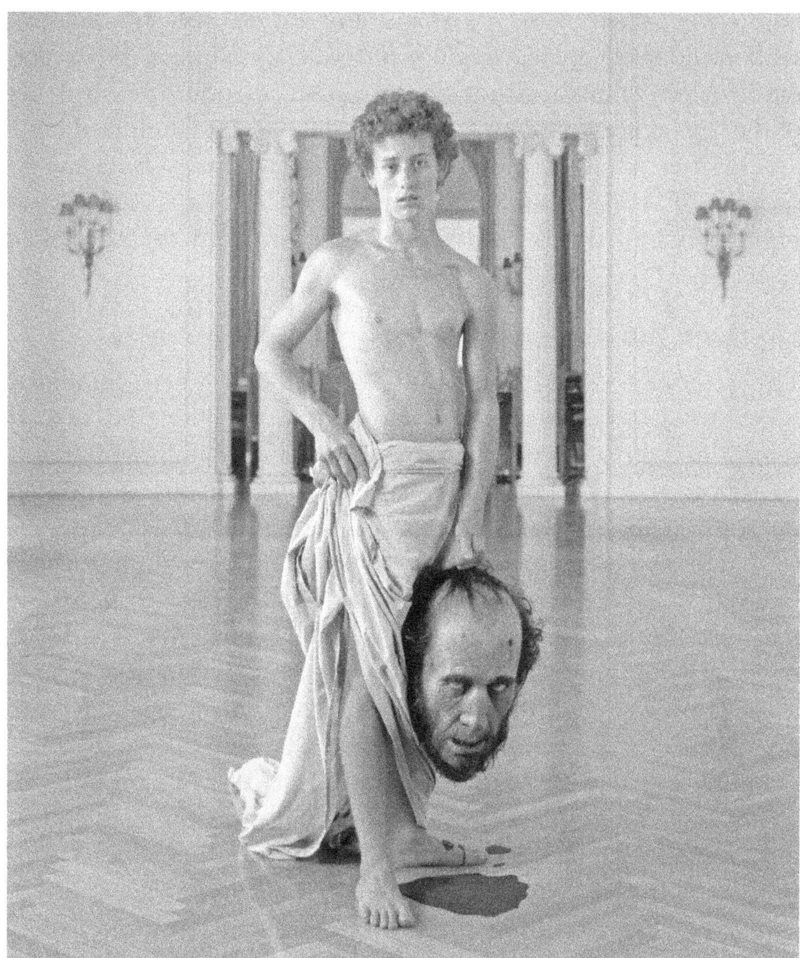

Figure 3.3 Charlie White, *Champion*, 2005.

Artists such as Jeff Wall and Gregory Crewdson use digital technology to assist and extend the processes they used prior to the digital era to represent their version of constructed and heightened realism. Others such as Charlie White step beyond the periphery of realism into the phantasmagorical to create images which play with illusion and the illogical.[23] These artists and their working methods reveal some of the possibilities the digital era enables. They demonstrate the potential for presenting highly convincing constructs. Their work extirpates the divide between physical reality and non-physical reality; the polarity between inner and outer-space loosens and they come together, an indicative construct of the digital era. As Edgar Allen Poe asked over two hundred years ago, on how we might distinguish reality from fantasy, 'Is *all* that we see or seem, but a dream within a dream?'[24] Or is it that with digital photomedia all that we see is one reality of many other possible realities?

Light up: the screen space of digital photomedia

The screen is significant in the digital era; when combined with increasingly integrated computing technology and display possibilities it is a powerful conduit for information dissemination.

Photomedia works are now presented on luminous screens that are sophisticated and portable. The screen is almost as ubiquitous as digital photomedia; viewing locations and opportunities have multiplied as the screen finds a position in every conceivable location. The projection of the digital environment in screen space continues the shift from the traditional physical connections, such as light projected through plastic film onto a screen, that were prevalent in a purely analogue environment, into an increasingly light-based electronic environment. The image is formed by light, space and time – it is transmitted or presented as electromagnetic information on luminous screens – all light-based operations. This process reaffirms the supremacy of light as a constant characteristic of photomedia and the digital era. The instantaneous transmissibility of light-based information has contributed to a multiplicity of opportunities for viewing and engaging with photomedia as light.

Fundamental to the process of perception, the screen makes visible the invisible; it is the interface between information and image. This relationship between the screen and the viewer is one that is far-reaching, encompassing the physiological and psychological because it is part of the increasing and omnipresent image-space of inner and outer experience. I maintain that photomedia screen space inhabits and connects with personal, private space and public, external space in a Bachelardian loop of inner and outer-space which constantly form and inform each other.[25] As screen-based imagery creates this engagement with inner-space, even if for a moment, we carry with us a fragmented memory of this engagement.[26] However, this relationship to the screen is not new – early image display devices such as the magic lantern[27] and the Zoopraxiscope are antecedents of the technology we use today.

In Erkki Huhtamo's[28] essay, *Elements of Screenology*,[29] he claims that the growth of the screen as entertainment occurred concurrently with the introduction of magic lantern shows and larger scale projected screenings of the 1800s, a relationship that he claims arose with the 'emergence of the urban, technological media society'.[30] Engagement with the screen is not new, occurring in the era of early image machines, as Huhtamo contests. What is revolutionary is that screen spaces may now combine with the computing power of digital photomedia to proliferate light-space-time quickly and responsively. They too are becoming a ubiquitous component of the photomedia experience.

At the speed of light

Denis Beaubois' *The fall from Raiatea* (fig. 3.4), a multi-screen video installation, was made as part of the artist's ongoing enquiry into the technologising of vision and its effects on perception. The installation is a configuration of large, wall-mounted screens displaying video footage captured from the transmission made from a cluster of small, wireless video cameras.[31] The footage shows the moments in which the cluster of cameras are physically dropped from the twenty-seventh floor of one of Australia's most dominating modernist buildings, a Housing Commission tower located in the Sydney suburb of Waterloo. A momentary sense of beauty that is visible from such a height transforms into anxiety as the cameras fall. As they spiral down from the building, the image continually breaks up and reforms as the ground approaches with increasing speed and increasing anxiety as the inevitable impact looms ahead. As the cameras drop, the action of falling is physically felt and symbolically implied; in this case it is the collapse of modernist ideals and the utopia of technological movement toward a better way of living.[32]

Most of Beaubois' cameras descend into destruction, but one camera breaks up on impact, stutters for a moment then flickers and blinks back into optical action, the camera vision is tilted and stagnant in the aftermath. Transmission eventually stops and visual

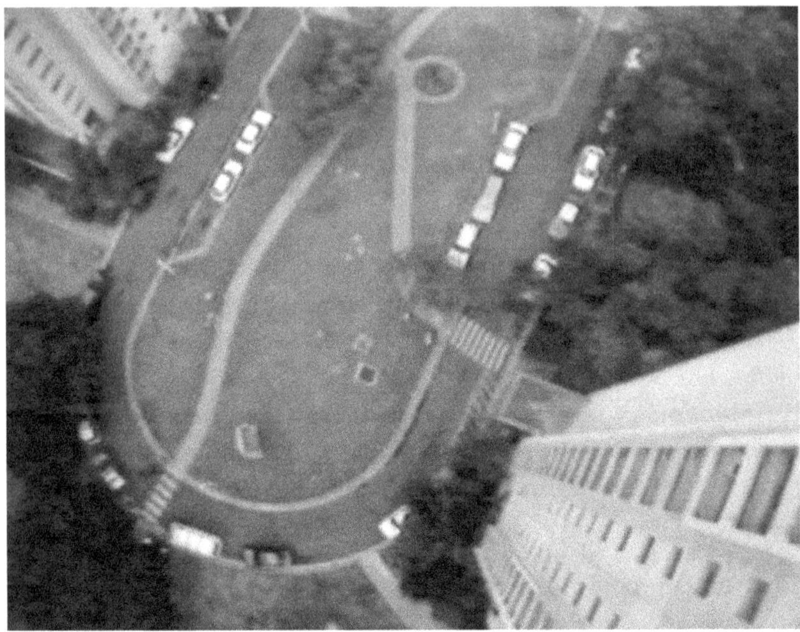

Figure 3.4 Denis Beaubois, *The fall from Raiatea*, 2007.

awareness is reduced to television snow. Beaubois states, 'The work explores the links between remote viewing and physical reality (passive consumption and experience).'[33] Through the cluster of cameras one may remotely view the event, which is physically felt through the experience of the action of dropping the cameras from a height.

In Beaubois' piece both the physicality of space and the durational rush of acceleration are played out through the perspective of the cameras as they simultaneously plunge and transmit. As Beaubois points out, instantaneous real time transmission has implications for the nature of vision and physical experience and understanding.

Slow motion: light-time in the digital era

The technological drive toward ever faster ways to take, send and receive the photomedia image has left us with light-space-time relationships and experiences particular to the digital era. The most unexpected of these is the emergence of slowness that comes with immense speed; this is not stillness as a polar opposite of speed as seen in the analogue era (see Chapter 2) but slowness as a result of speed, a paradoxical slowness.

While the fast transfer of light continues to be paramount, the still image holds its position as a totemic indicator of events, solidifying the transient moment. As Milan Kundera points out in *Slowness*, 'there is a secret bond between slowness and memory, between speed and forgetting.'[34] In an epoch where the speedy transmission of images proliferates, this fast succession of information leaves little time for contemplation and even less for memory. Hence, there is a certain appeal in stepping aside from this torrent of visual imagery to reclaim reflective time or the time necessary to consider and distil what is seen – to embrace the slow and the still. Perhaps this desire stems from nostalgia, or resistance, or, more likely, it is a balancing act where the slow is embraced as a way to bring equilibrium to the fast. David Campany theorises that in some contemporary photography which he terms 'late photography',[35] slowness is a dissociative stance. He states that, 'in forfeiting any immediate relation to the event and taking up a slower relation to time, "late" photographs appear to separate themselves from the constant visual bit stream emitted by the convergence of modern electronic image technologies.'[36] The viewer is no longer immersed in the specular but steps aside from the event to look at it from a temporal distance, and over time.

Video artists are addressing this slowness, and its correlative stillness and movement. Gillian Wearing's *Snapshot* (fig. 3.5) examines the iconic role of the photograph and its connection with human emotion through our collective desire to document and signpost life with photographic evidence of our existence, as was initiated with the advent of Kodak's Box Brownie (discussed in Chapter 2). *Snapshot*[37] animates the traditional photographic portrait by using long video takes of subjects who maintain steady poses.

The work consists of seven screens each showing what initially appears to be a still image, similar to those found in a family photo album. On closer inspection these 'stills' are, in fact, animated. Over the seven 'snapshots' the light-space-time of these moments is extended as if Bresson's 'decisive moment' has been stretched

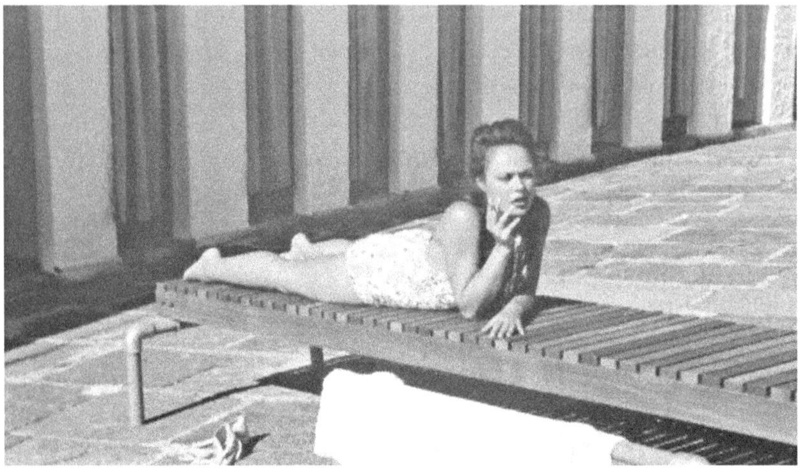

Figure 3.5 Gillian Wearing, *Snapshot*, 2005.

over minutes. Time appears suspended, durational; the image does not bear the usual marks of having been slowed down but instead focuses on a 'still moment' in the subjects' lives. Wearing asks her subjects and her viewers to slow down, to linger on a moment now gone, but perpetually present through the image-space of the snapshot. She allows the moment captured to hover and to shift.

In these seven different screen scenes, Wearing examines the historical trajectory of portraiture and the photographic record – from the studio portrait, to the Box Brownie, video and finally digital, re-staging her images to appear as if they carry the unique markers and signifiers typical of photomedia. She mimics the range of styles from the faded brown and white images similar to those of early studio portraits through to a glossy full colour image of a grandmotherly figure in a casual and contemporary living room. Across these different portraiture styles, one aspect is clear: despite the technological changes, the digital camera is used to capture and retain a part of an individual's personal history and experiences as image machines have always done. Over the series of images, Wearing displays a chronology; we move from the youngest subject to the oldest and from early photographic techniques and technology to the most recent. Each character encapsulates an expression of the varying moments of one's lifespan. One of the subjects in *Snapshot* suns herself, moving slowly to occasionally ash her cigarette. Is she lost in thought, or is she observing something beyond the picture frame? Most of the subjects in *Snapshot* look to a place beyond the image. Perhaps Wearing used this to suggest that something lies beyond the frame and beyond the light-space-time of the photomedia image-space. Is she telling us that these moments that exist and are photographed have longer histories and cannot be pinned down into a contained space or time?

David Campany points out that 'in the popular culture of mass media, the frozen image is often used as a simple signifier of the memorable, as if there were a straightforward connection between the functions of memory and the "freezing" capabilities of the still camera.'[38] In Wearing's work we are not given the entire satisfaction of a still image complete with its contained sensibility. These are not the embalmed moments that Roland Barthes discusses in *Camera Lucida*.[39] Barthes analyses the personal meaning of a photograph and the objective symbolic aspects of the image through a photograph of his mother as a child. The personal meaning of the image he terms the Punctum; the 'wound' the image manifests in the viewer due to their psychological relationship to the content. The objective meaning of the image he terms the Studium, the cultural and linguistic interpretation of the photograph. Importantly Barthes theorises the relationship between photography and time and considers that the still image embalms a moment in time and in a sense halts life, preserving it through the continuum of time so that the photograph becomes an image of life after death. Instead, in Wearing's *Snapshot* the 'moving still' leaves the viewer on edge; past, present and future come together and we are left constantly renegotiating and contemplating the essential nature of time and its spatial arrangement in memory and photomedia. These 'moving stills' flow, they endure and shift, like image-worlds.

Likewise, since the late 1970s, Bill Viola has exclusively used video technology as a way to explore the human condition. Viola's thematic explorations of human emotion, perception and understanding have been expressed through repeated use of the physical body to inform the analysis of space and time components

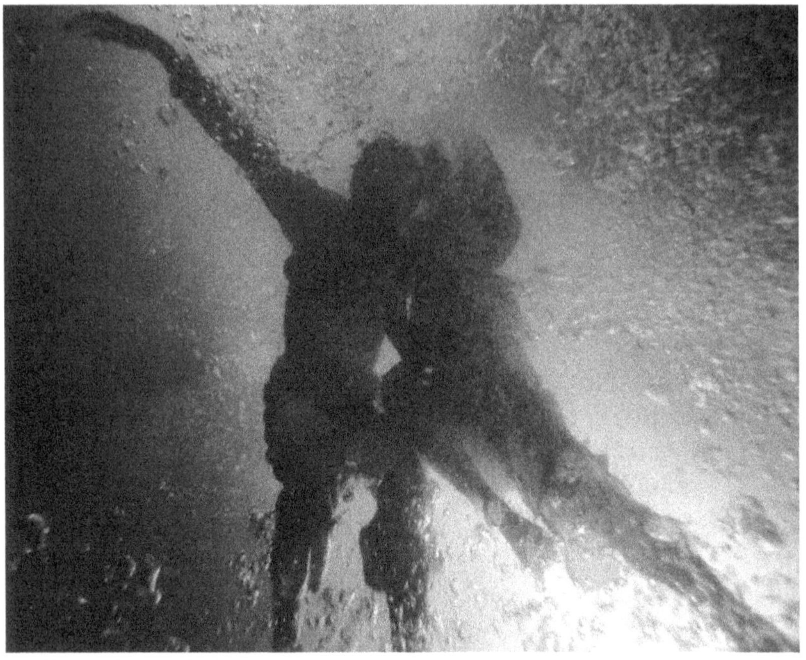

Figure 3.6 Bill Viola, *The fall into paradise*, 2005.

of video technology. These elements are used to profound effect in Viola's *The fall into paradise* (fig. 3.6). The narrative basis of this work is the medieval fable of Tristan and the unique love of Tristan and Isolde which was not able to be expressed in life, but only through transcending their physical form through death. In the course of *The fall into paradise* they move from this material world of physical presence, ascending or 'falling into paradise' and incorporeality.[40]

Viola's *The fall into paradise* begins with a tiny spot of light. The particle of light slowly and silently grows, seeming to move closer, suggestive of a single photon or even the pinhole of light that can be seen inside a camera obscura. Slowly, appearing ever closer, the dot of light expands. In these transitional moments the phenomenological experience of time seems very slow. The room in which the installation is located is veiled in darkness; everything beyond the focus of this emerging particle of light appears dense with this darkness. Ocular awareness dominates the senses with an intensity that builds as the moments progress. The room remains silent, only the viewer's own internal bodily sounds such as breathing can be heard and felt. Gradually, the speck of light forms into what appears to be physical bodies: at first only an arm then, as the movement of the body increases, one person becomes two and the two figures come closer as if moving head first in slow motion. This moment of understanding what one is looking at is immediately broken as the two protagonists, Tristan and Isolde, burst through an invisible divide between one space of air (symbolically their corporeal reality) and another of water (their spiritual selves).[41] At the same time we hear the sound of bodies plunging into water. Above perspective shifts to below, forward shifts into reverse as the two figures plummet into the water. Viola's use of prolonged slow motion is accentuated as the sounds of water from below the surface seem to engulf and submerge the installation space. The image slowly darkens as Tristan and Isolde continue their descent – or is it ascent? – into paradise; soon they fade away and the screen returns to darkness, the room silent again.

Engagement with this work allows one to enter phenomenological time, where reflection on the moment deepens through the unfolding detail of each seemingly unchanging moment. This specific connection to time is reminiscent of the images of Muybridge, Marey and Edgerton in that the properties of the photographic medium are used to reveal the detail of every moment. *The fall into paradise* is both analytical and expressive. Time is broken down but not through a discrete and deconstructed view of motion-through-time that chronophotographic studies provide. Rather, Viola's work uses photomedia to reveal the expanse of expressive moments that are lost in real time but are revealed in extreme detail when played out in slow motion. This presentation of time is akin to the subjective time one might experience in an intense situation such as that felt by Tristan and Isolde, when time seems to slow down and perception heightens to receive every detail of each moment. As the viewer lingers with *The fall into paradise*, one becomes aware of the durational experience of time; each moment although apparently similar is in fact, different. With this work Viola successfully heightens the experience of visual perception while arousing a sensation of the passage of time and the constant desire to embrace, hold and cognitively understand time while it slips away.

Similarities between *The fall into paradise* and Yves Klein's *Leap into the void* are clear: each is suggestive of the limitless possibilities of the void. Klein's analogue exploration of the void is realised again and expanded by Viola in the era of the digital image machines. Viola's work is an exploration of the nature of experience in this environment; the paradise that Tristan and Isolde descend or ascend into is representative of what is unborn, a void. As discussed in Chapter 2, this is similar to Flusser's concept of the void. In *The fall into paradise* Viola draws on his personal experience of Zen Buddhism. The void, as understood in Buddhist philosophy, is an absolute truth where 'appearances are like a magician's illusions; they are actually devoid of intrinsic reality.'[42] In this sense the void is all existence; an expression of the endless possibilities contained in one's perception of reality. This understanding in turn discards any single reality from existing as the only true version of the real. In both Viola's and Klein's work photomedia technology captures the tangible but suggests the invisible void. It is this void, or no-place, which comes to represent these forces that affect us increasingly more than the tangible. Through the visual photomedia surface we may go deeper toward understanding the depth of our own perceptions and the limitless possibilities before us in constructing our own reality, in constructing image-spaces.

So, as the flow of the image data stream continues to gain momentum it is with great interest that I bear witness to the progressive dialogue between the still and the moving image as explored through the video work of Bill Viola and Gillian Wearing among others. Such work is positioned between the structural definitions of photography and video. These works deliberately do not contain the often used freeze frame, instead they display characteristics of both still and moving images and in doing so connect with a dialogue concerning light-time, specific to the digital era. It is an era where speed is paramount and instantaneity is omnipresent. An era in which the slowing of time is a contrary response to the contemporary experience of speed. In Wearing's work the entirety of the image is instantly visible yet requires the durational commitment of the viewer to reveal the delicate nuances between moments to examine the time based changes. Each piece engages the subject in a traditional study through a fixed camera position; a technique which places the viewer in a conventional relationship to the subject, and one that has been used since the advent of the medium. These artists have used the technology to extend the moment, to play with a sense of time. In doing so, each work alerts the senses to the deep time of the moment, revealing what occurs when the relationship between the still and the moving image is played out as interchangeable via the specific characteristics of the video medium.

Void space: Digital photomedia and the loss of physical reality

'From today photography is dead'[43] was a much quoted but unreferenced declaration made at the time of the introduction of digital photomedia. Such a statement might arise from an incorrect understanding of the properties and function of photomedia, based on the belief that the image is connected to the object it represents in an indexical way. As Geoffrey Batchen states, 'photography's plausibility has always rested on the uniqueness of its indexical relation to the world it images, a relation that is regarded as fundamental to its operations as a system of representation.'[44] In the digital era, where tangible objects or the physical world at large are no longer necessary for their inserted presence into the photomedia image, a simple 'logic' persists which equates this new scenario to a loss of the real and ultimately a loss of photography. As Mary Anne Doane claims, 'The promise of indexicality is, in effect, the promise of the rematerialization of time – the restoration of a continuum of space in photography, of time in the cinema.'[45]

The digital era erases this pledge of indexicality by making it possible to create an image without the need for a primary physical existence. Digital photomedia no longer requires a tangible object in order to present the seemingly real, thus removing any clear connection or indexical relationship to the real. The instability of reality and what signifies the real are made apparent and thereby generate an anxiety concerning the fundamental instability of objects and their place within the world. Instead of relying on Batchen's comfortable position on the uniqueness of photomedia's representation of the physical, we would do better to recognise that photomedia is light and light is the substance of all visual perception; it makes what we understand to be visual reality. The assumption that the physical and the real are mutually interdependent has lead Baudrillard, Flusser and Virilio to consider that in this new digital era all connections to the physical and therefore the actual are lost and replaced by a new space of nothing, a void. This is something that Viola and Klein explore through their own analyses of the void. However, this is not an era of post-photography as Geoffrey Batchen proclaimed in a 1994 edition of *Aperture*.[46] What Batchen failed to comprehend in that journal article is that photography's past, present and future concerns its inherent connection to light, therefore this is not the era of post-photography but rather it is an evolving configuration of light-space-time structures. The digital era is a new but contiguous era in the history of photomedia characterised by distinct light-space-time structures and experiential possibilities.

It seems that the void is a recurring theme consistent in both Eastern and Western thought. As discussed, in Eastern philosophy this notion of the void is prevalent in Buddhism and embraces an all encompassing notion of the potentiality of all realities at any time. While in Western thought the void began as a construct of the ancient Greek philosopher and 'atomist', Democritus,[47] who considered that all eternal things were composed of particles floating in a space, which he described as 'the void', 'nothing' and 'the infinite'.[48] This dichotomy of 'something' and 'nothing' persists throughout Western philosophy and becomes a significant perspective of the digital era.

As Baudrillard claims, 'everything pivots upon the art of disappearance. But nevertheless, this process of disappearing has to leave some kind of trace.'[49] This concept of a trace is suggestive of an outline of the event that remains after it has taken place, such as an image. Further, Baudrillard argues that in the world of images, 'events no longer generate information';[50] instead they are reconfigured 'into a present/current void where only a visual psychodrama of information'[51] remains. This visual trail is the trace which comes after the event and consists of reconfigured visual images that maintain no connection to the event itself. It is an illusion of a reality on the surface of the image which is the trace. As he says, 'the photographic image materially translates the absence of reality'[52] in that it no longer references 'reality' but instead, the 'trail of images' reflected on the screen. In this place, or perhaps 'no-place' of lost space and time, disappearance and displacement, an image-space unique to the era is constructed through fragments, motifs and hallucinatory scenes generated by the plethora of images which are the self-referential traces of prior images.

Paul Virilio connects the proliferation of the photomedia image to an increase in speed; he considers that if speed is light – all the light of the world – then what is visible derives from both what moves and the appearances of momentary transparencies and illusions. Virilio's theories of 'dromology' or the 'logic of speed'[53] lead him to claim that all we are left with is the speed of light.[54] The digital era has accelerated to this constant and produced an instantaneous transmission of images. Daguerre's aim with the invention of photography was to attempt to instantaneously capture the spontaneous action of light. While the analogue era came close to achieving Daguerre's intention, it is not until the digital era that light-time is definitively characterised by instantaneity. Aspects of the instantaneous image were witnessed in the analogue era with the introduction of television and video. But, in the digital era instantaneity is a consistent characteristic enabled via fast data transfers of digital information which occur at the speed of light and are potentially connected to any place, any time.

Images are no longer captured and stored as latent information to be developed; instead they are transmitted as data to the viewer, readable via any number of screens as the event unfolds. For Virilio this epoch faces a dromological scenario, where at absolute speed – the speed of light – we reach the void, and 'the end of the physical world as a dimensional truth'.[55] At the speed of light the lag time between an event and our knowledge of it is instantaneous – the physical gap between real time and 'image' time implodes and collapses geophysical location. At such a speed we, as Virilio would have it, are involved in 'the ultimate traffic accident where at the speed of light the apparent reality of the visible world comes to an end, implosion, dimensional collapsing [*télescopage*] that would see the disappearance of appearances in the dazzling light of speed'.[56] Virilio considers that through the action of light and the excess of speed that results, the viewer enters into the void. Passing into the void means moving into a new space, a light-space that is beyond geophysical location, which is a no-place. This is similar to the void as discussed by Flusser, Zen Buddhism and Klein.

At the crux of Baudrillard's and Virilio's concept of the void is the dubious relationship between the real and the image. Arthur Kroker states:

> When reality is exposed as simulation, theory as artifice, the sign as terror, and bodies as only apparent perspectives, then we can finally know that Baudrillard's thought had about it that certain pataphysical quality of always descending to the height of the void; always, as Virilio would say, "falling upwards" into the desert of the real.[57]

What is apparent is that any attempt to provide a finite analysis of contemporary photomedia structures, let alone their experiential manifestation and connections to the real, is counter intuitive. As Kroker critically states when referring to the theories of Baudrillard and Virilio, they exhibit a pataphysical quality, a nonsensical pseudo philosophy which at first may appear to be a sound philosophy but in which, once properly analysed, a fundamental philosophical argument is not given. Ultimately, this digital photomedia exposes the fragility of any absolutism.

Whereas both Baudrillard and Virilio are concerned with the photographic image, in contrast, Flusser emphasises the affect of photographic technology, itself, through his philosophy of the apparatus.[58] Flusser, however, theorises that under the spell of the apparatus we move deep within the 'photographic universe', gradually losing sight of our origins. Similar to an image-space in that it consists of the complete output of photomedia, this 'photographic universe' is no longer connected to the physical world. Instead, this space is constructed purely from images, which are encoded by the apparatus of photomedia according to its programme. These images are then projected back into the void-like space of the 'photographic universe'. This is a scenario where perpetual improvement of photomedia technology induces what Flusser refers to as a 'coded' state of being where our actions are informed by photomedia and are harnessed in the service of continually improving it. In doing so society strengthens and contributes to the evolving 'photographic universe'. Flusser's point is that the increasing proliferation of photographs allows the domination of the apparatus to persist and so moves us further away from 'traditional' images and the physical world. Instead, we move closer to a cosmos composed of 'technical' images.

While for Baudrillard a void arises through the photograph which is an illusion of an object, 'the moment in which the world or the object vanish into the image';[59] Flusser's void is an image-space in which the distinction between the physical world and the image is lost due to the imperative posed by the technology to continually improve its capacity to encode images. The logical result of this scenario sees photomedia as an increasingly dominating system of technology-driven actions that produce ever more images which in turn spurs on the development of photomedia.

While the concept of a void does not inherently signal a loss of the real, it seems to do so in the philosophies of Baudrillard, Virilio and Flusser. As photomedia becomes ubiquitous, and if one persists in associating the image with 'reality', a sensation of loss may occur. Loss has been a persistent theme in photomedia discourse throughout much of the late twentieth century, which positions the photographic image as either referent or simulacrum.[60] As referent the photograph bears a faithful representation or connection to that which it is derived from. As simulacrum the photograph bears only a simulated reference to that from which it was derived. The image as simulacrum makes one's experience of photomedia a simulation of the real without any necessary connection to the real. In response to this Melissa Miles urges an analysis of the implication of light within digital photographic theory in order to 'Shift discourse away from tired debates about indexicality and representation, and toward an understanding of digital photographs as the products of multifarious arrangements of mediated, symbolic, material, technological, social, historical and discursive elements.'[61]

But the discussions outlined by Miles are well worn debates; let us instead move to image-spaces and their intrinsic light-space-time formations.

The digital image-space: a matter of light-space-time

The fragmented, fractured and always shifting formations of image-space are transmitted and received at a greater speed in the digital era. Due to the incessant bombardment of images fragmentation occurs not only in the public realm of the images but also in the inner-space of understanding. As Baudrillard states: 'contemplation is impossible, images fragment perception into successive sequences and stimuli to which the only response is an instantaneous yes or no – reaction time is maximally reduced.'[62]

Image transmission has reached such a speed that there is no time to think, there is only a 'knee jerk' response to the barrage of image stimuli. Such a system of communication induces an immediate response of 'I accept' or 'I reject' the image. This experience of uninterrupted images fragments inner-space and thus understandings are altered as a direct result of this experience.

Fragmentation is also a feature of Virilio's philosophy of the photomedia image. In his 1988 text *Pure War*, Virilio discusses the characteristics of the era as a fragmentation of inner (personal) and outer (public) psychological spaces. Virilio speaks of interruption, not continuation as constituting experience in this era. He states: 'Our vision is a montage, a montage of temporalities which are the product not only of the powers that be, but of the technologies that organise time.'[63]

In a recent publication, *The Negative Horizon: An Essay on Dromoscopy*,[64] Virilio cites Benoit Mandelbrot's[65] text, *Objets Fractals*, in which Mandelbrot claims spatial dimensions are hardly more than fragmentary messages that geometry will never cease interpreting.[66] That is, we never cease to try to make a coherent 'space' out of these fragments which have no other pattern than that which is ascribed to them; geometric interpretation of their forms configures a structure similar to that found in the fractal geometry of nature as theorised by Mandelbrot.[67] This interpretive configuration similar to the structure of image-space allows a reading of each fragment (the micro) in relation to other fragments within a larger system (the macro). That is, each image, which in itself can be further fragmented, is identical in pattern to the larger system known as image-space. What is identical in either the perceived fragment or the whole is that they are light-space-time.

Conversely, the model of photography Flusser puts forward allows for a reading of the 'photographic universe' as 'a model of ourselves, of the world and society'.[68] By this he means that it is we who have been involved in its formation, we who have 'imaged' our world in photographs, even if it is under the spell of the apparatus. Flusser is not concerned with the prismatic structure of his 'photographic universe', instead he urges one to break free from the 'magic spell' of the apparatus or rather the photographic system. He states:

> the task of a philosophy of photography is to reflect upon this possibility of freedom – and thus significance – in a world dominated by apparatuses; to reflect upon the way in which, despite everything, it is possible for human beings to give significance to their lives in face of the chance necessity of death. Such a philosophy is necessary because it is the only form of revolution open to us.[69]

Flusser offers not just a philosophy but a call to resist the control of the apparatus and to bring about one's significance and importance before the inevitable death we universally face. He is arguing for a level of consciousness within photographic practice that will enable a revolution. Flusser's philosophy is, in short, that there is no freedom in a world of automated, programmed and programming apparatus. But, through a philosophy of photography, an understanding of all the aspects of photography from the camera to the image to the programs in which the apparatus encodes behaviour, unexpected photography may emerge that breaks free of this historical paradigm.

One strategy to 'freedom' is collage. This is a technique artists use to construct new image-spaces and hence understanding. Collage creates convincing experiential structures in which to navigate and consider the relationship we maintain to the world and to photomedia light-space-time structures.

David Noonan and Andro Wekua are two examples of artists who tap into the possibility of escape from the thrall of the 'apparatus'. They do this by sourcing images from the existing image-spaces and reconfiguring them to construct an alternate space.[70] Noonan and Wekua do not exclusively or necessarily use digital technologies for the production of their work, however, their work is demonstrative of the digital era and its image-spaces.

David Noonan creates collages, video montages and installation environments. A feature of his work includes the persistent use of recurring motifs such as owls, Tudor tropes[71] and theatrical figures sourced from 1950s and 1960s theatre periodicals and books. These elements are brought together in a hallucinatory pastiche; the resulting imagery acts to displace the viewer in image time and space. As Jennifer Higgie claims, 'these images – ready-mades reassembled – hint at archaeology; they mine time via its residue.'[72]

In *Kabarett Keif* (fig. 3.7) Tudor era visual motifs occupy the same image-space as a man with a striped shirt, fez and bra reminiscent of a clown in the circus. Within the picture plane other men look on at the scene from beneath the many layers of the image, while the central figure looks directly at the viewer. The floating symbols position the viewer firmly within the imagery and in dialogue with their personal associations and ascribed meanings of the scene. Melissa Gronlund considers Noonan's work to 'explore a state of what might be called temporal exile: hangovers from the past in the present'.[73] This nostalgic reframing of time and space is indicative of the era; it is an 'enduring light that contrasts with the dark... an intimacy [is] created by bringing representations of the past, things that happened "before", into the "now".'[74] This method employed by Noonan follows Flusser's appeal for freedom from the apparatus. Noonan does not take images according to the program, he takes existing images, fragments them further and recombines them, creating new meanings, and by doing so restructures the flow of the image-space.

Andro Wekua also uses collage to create images which are suggestive rather than didactic. His work, like Noonan's, spans collage and sculpture within an installation environment, and demonstrates the fragmented nature of the digital epoch's light-space. Also like Noonan, Wekua's work is derived not only from the deluge of images located in image-space but from his personal memory and connections with particular motifs that he continually re-uses. But, unlike Noonan, Wekua draws on a history which

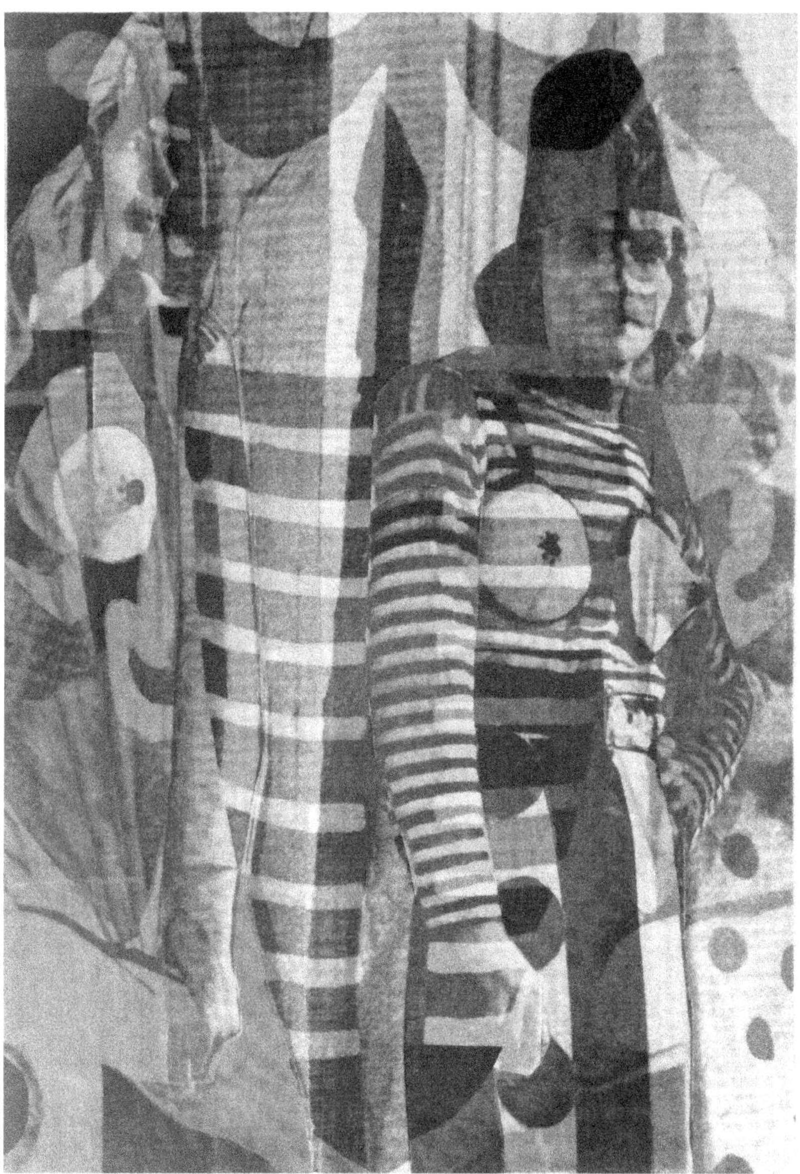

Figure 3.7 David Noonan, *Kabarett Keif*, 2007.

originates in Georgia, by the Black Sea; a history of persecution and displacement. However, what is memory and what is fiction in Wekua's work is not clear. These themes seem to defy any specific narrative or reference.

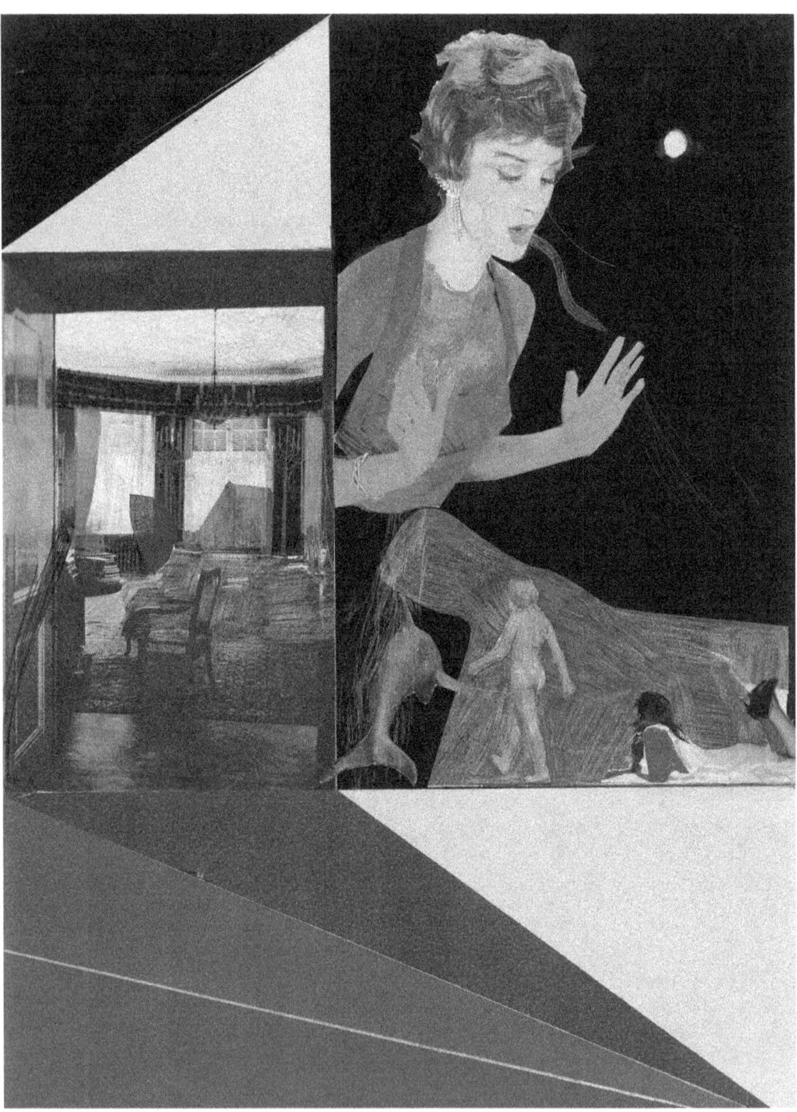

Figure 3.8 Andro Wekua, *Lying, Walking, Swimming*, 2005.

Lying, Walking, Swimming (fig. 3.8) constructs a fragile reality formed equally by ambiguity and tangibility. Gestures, such as that of the central figure, only confuse – is she bending forward to take the breath of life or give it? A small sun or moon is nearby, ineffectually watching. An empty room fractures the picture's plane of view; contributing to this effect, a dolphin is in the same perspective as a naked child, 'walking' as the title suggests. Scale and perspective are askew; strong colour blocks further fragment the image. Dieter Schwartz comments that such shapes, 'by means of their configuration, apparently abstract forms ... recreate the hidden connection that the memories have lost'.[75] Geometric shapes and colour represent the structure of Wekua's language, a visual language which uses both abstract and figurative tropes to express an abstruse narrative. This structuring of the images reconfigures the associations Wekua has to them, and at the same time conveys new relationships for the viewer to interpret. 'Wekua forms a new reality. His protagonists are arranged by means of the special framing virtually as on a stage.'[76] And this is how he ushers the viewer through meaning in his work. As Sally O'Reilly explains, 'Much contemporary collage draws on the symbolic and metaphorical potential of illogical combination, radical subjectivity and fantastical self-governance.'[77]

Indeed, collage is contrived by the maker and interpreted by the viewer. O'Reilly goes on to state: 'and just as the initial proliferation of collage coincided with the establishment of the telecommunications and transport systems that would lead to the non-linear and exponential expansion of today's information networks, now collage is a succinct reflection of their effect.'[78]

Collage takes fragments from the ceaselessly forming and re-forming image-spaces, and reworks the fragmented connections between images and meanings through personal, illogical or dream-like juxtapositions. This activity is progressed further by artists such as Lothar Hempel, Sara VanderBeek and Tom Burr who conflate image and object polarities through their contemporary forms of assemblage. Their methodology for working within contemporary photomedia environments and the ensuing image-worlds features an arrangement of light-space-time structures that are uniquely ordered and mediated through the individual who constructs them.

Tom Burr, through his art practice, references a fragmented trans-historical idea of the twentieth century artistic subject through objects and images. Burr relates his practice to the vinyl record, which is an interesting and useful analogy for artistic interpretations of light-space-time structures of the contemporary era. He claims that:

> the record became interesting to me, it does have a beginning and an end but it is endlessly repeated, so there is some sort of an end I suppose but now it's simply a recorded fact that is open to interpretation and memory and loss of memory that we repeat fragments of.[79]

In this sense Burr, like other contemporary artists, is taking the recorded image and engaging with his personal relationship to it for illogical and personal compositions that play against repetition and the Flusserian model of control held against the human by the apparatus.

Daniel Crooks has an interesting perspective on the contemporary experience of photomedia structures that is less physical but equally fragmented. He employs slit-scan techniques which have been used prominently in the scientific field since the mid-twentieth century and in visual arts since the 1980s by artists including Pipilotto Rist with *I'm Not the Girl Who Misses Much* (1986), David Tinapple with *Slit-scan* (2001) and Charles Broskoski's *Infinity x 10* (2006).[80] One of the most well known examples of this technique is the 'stargate' scene from Stanley Kubrick's film *2001: A Space Odyssey* (1968),[81] which shows the central character Dr. Dave Bowman moving though an abstract field of spectral light rays. A more recent example is Andy and Larry Wachowski's film *The Matrix* (1999).

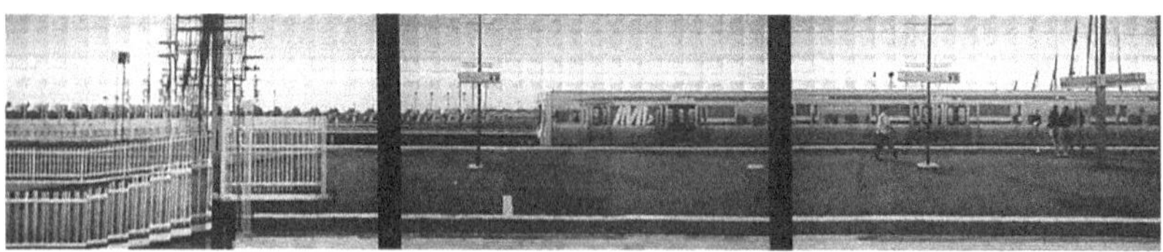

Figure 3.9 Daniel Crooks, *Train No.1*, 2002–2005.

Slit-scan is a photomedia technique which involves a moveable slit placed between the camera and the scene, which has the effect of recording a horizontal or vertical slice of the larger image. When each sliver of time is recombined into an image the result is an image which contains slices of light-time from different moments that occurred during the recording which, stretched across a screen, produce a new light-space perspective. In the digital era the physical slit is not necessary as the computer now allows the collection or filtration of visual information taken from the recorded image processed to show a slit of this image. Crooks applies this technique, which he calls 'Time-Slice',[82] when analysing movement through time by shooting, for instance, his local transportation system, the Melbourne train line, or an athlete in training. Projected across three large images, *Train No.1* (fig. 3.9) begins in stillness then moves to a fragmented view taken from a station platform of trains and people moving past. Then, the camera, along with the train it is in, moves along the train line, and as the platform and the usual signifiers of an urban landscape such as roads, electricity poles and tracts of scrubby industrialised land go past, they 'warp and ripple in their own segments of time'.[83] Through the slit-scan technique layers of vertical slits arranged horizontally are spread across these three projections and, as *Train No. 1* progresses, a time-shift occurs in each of the slits, having the effect of stretching time in a fragmented and distorted way.

The connection between Crook's work and Marey's is obvious; through photomedia, both achieve a dissection of time and space in the analysis of movement. What is not so clear, perhaps, is the connection Crook's work has to space and time structures of the digital era, in which the rigid segmentation and division of time is precise. This precision is in counterpoint to the free flow that occurs visually in Crook's work when each part of the image flows from one fragment to the next. Crook's work is suggestive of the experience of a multifaceted and fragmented experiential reality, which Crook refers to as a 'Polyocular' viewpoint.[84] As Laurence Simmons explains:

> Polyocular is a term that describes the small differences between images obtained from many different angles that enable our brain to compute invisible mental coordinates. It is also a concept used in

> organisational theory and in anthropology to describe our ability to understand cultures from more than one point of view.[85]

In this aspect it reprises the work of Duchamp and the Cubists who took pictorial space-time representations, broke them up and re-assembled them to conform to multiple simultaneous points of view. Like these twentieth-century visionaries, 'Crooks seems to question what other space-time continuums exist'[86] beyond those immediately apparent.

Noonan, Wekua and Crooks construct new collaged image-spaces which play against the Flusserian apparatus. By fragmenting and reconfiguring collected imagery from surrounding image-spaces, the unexpected occurs and freedom is gained by playing against the normal processes of the apparatus.

The digital era evolves from the larger history of photomedia but has maintained its own specific characteristics such as heightened fragmentation of image-spaces, and the instantaneous speed of light-time and its antithetical expression in slowness. But, as revolutionary as the digital era may seem as it launches us into the void, there is yet more to come.

4

FUTURE IMAGE MACHINES 2039: TWO HUNDRED YEARS AFTER THE INVENTION OF PHOTOGRAPHY

The future?

When thinking about the future what does one say, think, imagine or believe it will be like? If considering the short term future of tomorrow, one might see a future much like the current situation. But when considering the distant future, 20, 30 or a hundred years from now, a substantially different scenario undoubtedly arises. A book I read as a child, *The Usborne Book of The Future: A trip in time to the year 2000 and beyond,*[1] proposed that my future would include wearing leisure-suits while robots took care of the menial aspects of my existence; that I, and all other humans, would travel to other planets for personal enjoyment. Such propositions seemed attractive and plausible at the time. Now, as I revisit that book it is not the naivety of this futurology which strikes me, it is the paradigms that were used for constructing notions of the future, paradigms which continue to be used to construct scenarios of the future.

When postulating the future it is important to question whose it will be. The concept of a singular and universally shared future may be more absurd than any far reaching idea of what the future may hold. Overarching models of a technological future often involve a single homogenous culture. Zoë Sofia in her essay 'Contested Zones: Futurity and Technological Art'[2] claims that relative positions and views are an accepted aspect of understanding any history but that these same perspectives are not often applied to ideas concerning the future. That is, in Western culture, at least, the idea of the future is always considered in terms of a singular future as in 'the future' and 'assumed to be a destiny shared by all'.[3] We understand that this is not usually the case; the multifarious nature of societies and perception can also translate into multiple iterations of the future. The fragmented nature of experience does not necessarily stop in the future, it may continue; therefore we must consider numerous coexisting realities when speculating on the possibilities that may be manifested in the future.

Furthermore, Sofia claims that to speak of the future with certainty, as if things will or are likely to happen, is to speak in a 'collapsed future tense'. The problem with such 'future speak' is that by doing so we risk pre-determining the outcome and placing technology beyond the concerns of human decisions or actions that may be contested from other perspectives. In this scenario she claims we close ourselves off to the potentiality of the unknown. Sofia asks: 'Can a future already known in advance even be considered a future?'[4] This question highlights that claims about the future are in part considerations of one's current predicament, as well as statements that predestine the outcome.

In contrast, or in a way complementary to Sofia's argument, Yoko Ono explains when discussing her conceptual location '*Nutopia*', a place where we are all together as one, that:

> to say it exists on a conceptual level is to say that the country exists in all of our minds, and in our hearts, and that's very important to understand. Because first there's an idea, and then we imagine that idea as a reality. Through the imagination things do become reality – a physical reality. So it's not just a belief, but it's a recipe for creating reality. It's a recipe that works.[5]

Here Ono suggests that we may all create a future reality by conceiving it. Walter Benjamin made a similar claim when he stated that 'Every epoch not only dreams the next, but while dreaming impels it towards wakefulness.'[6] The implication is that in speculating on a possible future, we wittingly or unwittingly bring it into being.

Ono's and Benjamin's ideas that to conceive or dream of something creates the grounds for a future reality has a pertinent application to photomedia. In Vannevar Bush's 1945 utopian propaganda essay 'As We May Think'[7] he states 'certainly progress in photography is not going to stop.'[8] He comes close to predicting contemporary digital photography by claiming that a scene may be captured by a photocell.[9] Today, digital photomedia uses photocells in a cluster to read and translate light into a digital signal, such as a CCD. Another of Bush's speculative concepts was the 'Memex', an abbreviated term, coined by Bush, for 'memory extended'.[10] The 'Memex' was to be built into a desk with a keyboard, microphone and an array of screens much like the desk at which I write this book. Bush's vision of this machine-assisted memory used microfilm copies of photographs and papers and if one wished to add anything to the archive they could do so via a camera or a touch screen. When away from the 'Memex' the user could mount a camera on their forehead to capture more microfilm images. An important aspect of the 'Memex' is that Bush proposed that, 'with one grasp, it snaps instantly to the next that is suggested by the association of thought, in accordance with some intricate web of trails carried by the cells of the brain.'[11]

What Bush was envisioning was a proto-hypertext system[12] and today the 'Memex' is considered to have influenced the structure of the World Wide Web which uses hypertext for linking documents. Although we may not yet be able to materially track the rapid flow of associative mental concepts, we can with minimal physical effort navigate conceptual 'links' using the internet and a mouse and keyboard. A direct realisation of Bush's idea can be seen in the *MyLifeBits*[13] project of Gordon Bell who has turned his entire personal history – articles, books, letters, photographs – into digital records. Bell's project is based on a desire to fulfil Bush's concept of the 'Memex'. For Bell to realise his goal of going paperless, he required a personal assistant working solely on the project for several years. Now, as part of the *MyLifeBits* project, all telephone calls, web page visits, and transcripts of every instant message he sends and receives are recorded. With the Sensecam, a wearable camera that Bell uses each day to take photographs, a Global Positioning System (GPS) keeps tag of Bell's location and is attached to the image data. In this way, the *MyLifeBits* software automatically assigns the geophysical position and time each image is taken.

Many others have also wanted to capture every possible aspect of their life. For example, from 1915 to 1983, Buckminster Fuller worked on the Dymaxion chronofile[14] which involved Fuller documenting his life every 15 minutes in a scrapbook. Bush and Fuller are not alone in the attempt to chronicle their life through continuous time, they are joined by other lifeloggers such as Steve Man who has been documenting his life through live, 24×7 video streams since 1994. This desire to record and share many aspects of one's daily encounters is now common and is manifested online, where people actively document and share the minutiae of their life. What was once a utopian quest for the likes of Bush, Bell and Fuller has become a reality for millions of people today and is likely to continue into the future.

Nevertheless, predicting any aspect of the future with accuracy is challenging and often biased to one's own experiences and understanding. Utopian and dystopian perspectives are a fundamental aspect of human understanding, and influence perceptions of the future. Martin Lister suggests that while dystopian technological perspectives fear that the machine will come to dominate human civilisation, utopian perspectives conversely inflate the role technology plays in shaping the message it transmits,[15] as if the technology contributes significance and meaning to the message. Lister's point is that each perspective ascribes to technology a great capacity to effect experience. Lister goes further by claiming that what is perceived is 'A technology which can then be blamed or celebrated according to our predilection for the kinds of future it is seen to promise or threaten.'[16] In contrast, Flusser understands photomedia technology in terms of a system of communication that is not successfully interpreted or understood by society, and unless we 'escape' the program society will continue to be increasingly dominated by the 'photographic universe'.

According to the narrative of Western techno-historical development we will see an increasingly rich technological environment. Sophia argues this utopian narrative persists in the form of technological neophilia where everything new is good.[17] She suggests that this perspective sees a future where technology comes at us with great speed, from directions beyond our control.[18] Certainly, Virilio and Flusser would agree with this.

Apart from the extremes of utopian and dystopian paradigms, some predict change based on their understanding of the past; that the future is a continual progression of what is known, somewhat like image-worlds which constantly form and reform from what is currently available. Others speculate that at any given moment dramatic change to the growth and direction of any and all aspects of society, not just the technological sectors, could occur.[19] Also, claims of the death of photomedia are likely to continue. The photography historian Geoffrey Batchen theorises that in the future photography will not exist. He claims that 'photography's passing must necessarily entail the inscription of another way of seeing – and of being.'[20] Although Batchen does not elaborate on this claim, it is no doubt tied to his historical perspective of photography as a fixed technological tool predominantly tied to analogue processes. In view of this he forecasts that in the future we will inhabit a post-photographic world, where new visions and experiences, and a whole new way of inhabiting the world will emerge from new technological processes for photomedia. Unlike Batchen I do not see photography becoming obsolete because photomedia is not tied to the technology, it is tied to light and light will continue to be used as a messenger of image-based information. What will potentially change is not photography, but our understanding of it; we will come to understand more completely the role and importance of light within photomedia, its visual language and the changing light-space-time structures that we will experience. It will be difficult to ignore the importance of light as we reach the era in which photomedia images become an integrated part of understanding through the evolving image-worlds.

Bearing in mind the limitations of predicting the future, what might photomedia be like in 2039? As Jonathan Griffin suggests in his essay 'Future Conditional', 'in a culture swamped with dystopic images, perhaps it's time to resurrect the lost art of looking forward.'[21]

The photomedia technology of tomorrow

Photography is a technological tool, a technology of light, and one which is currently linked to the digital era. One method for speculating on how photomedia may function in 2039 is to analyse the growth of digital technology. Although it need not be tied to digital technology,[22] an understanding of this aspect of photomedia may assist in forecasting the direction of growth over the next 30 years.

Intel[23] co-founder Gordon E. Moore first observed and wrote on the projected growth of computer hardware in 1965. What is now called Moore's Law is generally understood to be the doubling of computing power every 18 months.[24] This is a trend that has persisted since Moore's initial analysis and is predicted to continue for some time yet. Currently, most theorists involved in any forecast of the future base their analysis on the exponential development of the technology[25] using Moore's Law.

However, futurist commentator and inventor Ray Kurzweil has projected one possible direction for the future of digital technology with his concept of the law of accelerating technological returns.[26] Kurzweil argues that overall capacity doubles every 1.2 years instead of Moore's 1.5 years. What this means is that in the twenty-first century we will not experience the same or similar rate of technological progress as we did last century. Instead, unlike the twentieth century, the equivalent of 20,000 years of technological progress will be experienced. Kurzweil theorises that this trend emerges because increased computing power has a multiplying effect on our knowledge, enabling greater research and hence accelerated technological developments. The human genome project, which mapped human DNA much faster than originally anticipated, is a case in point.[27]

Steve Dietz proposes in his essay 'Ten Dreams of Technology', ten common future predictions which are 'dreams of technology that have a future, even if we do not know it yet'.[28] They include symbiosis with technology, robotics, immersion, world peace, transparency of information and hacking into the mainframe. As wide and varied as these utopian or dystopian dreams are, depending on your perspective, they are just some of the many speculations that could potentially be developed to a substantial level in the future of 2039.

The implications of such advances involve a future where human reality as we know it ceases to exist. A complete paradigmatic shift results that sees human intelligence surpassed by machine intelligence until ultimately human civilisation reaches a 'singularity', an evolutionary stage where it is not possible to distinguish between human intelligence and machine intelligence as they merge, becoming one entity.[29] Kurzweil's singularity involves 'technological change so rapid it represents a rupture in the fabric of human history. The implications include the merger of biological and non-biological intelligence, immortal software-based humans, and ultra-high levels of intelligence that expand outward in the universe at the speed of light.'[30] This is a possibility that seems somewhat close to a concept of the void that Flusser and Virilio have discussed in that consciousness merges with the technology to such a point that they are inseparable and all encompassing.

If the prediction of the 'singularity' is plausible, then by 2039 the 'singularity' will be within reach and thus reality as we understand it today, and therefore photography, would become a light-space-time singularity.[31] This would result in the ultimate integration of inner- and outer-spaces, where the two become unified and form a singular entity, a universal space without the separation of inner and outer; the current distinction between reality and virtual reality will merge to form one cohesive light-space-time 'universe'.

Kurzweil and Moore offer us a proposition for the future which is based on continued technological growth. Even if this should not come to pass, a future based on increasing technological sophistication would undoubtedly encompass wide changes in the methods of shooting and creating photomedia images, as well as storage systems and display mechanisms.

Based on Moore's Law some general directions can be predicted and applied to the growth of photomedia technology of 2039. Overall, cameras will offer extremely high definition and three-dimensional moving and still recording capabilities.[32] Computational photography where the camera may be fitted with a small cheap lens but the computational power inside will produce detailed pictures wil be improved,[33] allowing portability and miniaturisation without sacrificing quality.[34] Embedded and mobile cameras will proliferate; ultimately the camera itself will reach the kind of ubiquity that the photographic image did in the analogue era.

Digital images will continue to be malleable quanta of data, able to be fixed in any space required through advanced display systems. Screen definition will be altered too, as brighter screens will make daylight viewing comfortable. Potentially, an increased range of objects will be implanted with self-powered wireless-enabled screen devices allowing the integration of photomedia images into everyday items, from large-scale objects like buildings and aeroplanes to smaller entities like jewellery and clothing.[35] These screens will be adaptable to three-dimensional surfaces and non-standard dimensions. This might be in the form of illumination devices such as Light Emitting Diode (LED) technology which is currently forecast to act as screens that will not only function on a similar range of supports as projection devices but will also be embedded in human skin.[36] The implication of this is that human bodies will also function as 'screens' for the display of photomedia.

Currently, photomedia can be captured and posted online and potentially viewed by millions of people, all in less than a few seconds. In 2039 this will be enhanced not only by the multiple screen environments but also through refined compression techniques and vastly improved internet speeds which will enable instant and ubiquitous display of photomedia imagery.[37] With these new screen environments and various touch technologies the image has the ability to function simultaneously as a transient packet of information and a tangible object residing in the touch screen. Thus, many more possibilities will have evolved for the way in which artists engage the viewer. Also, viewers may be contributors or collaborators to this content through interactive information sharing and authoring of online content, further expressing the inner and outer coalescence of photomedia image-space within screen space. Inner image-space and outer image-space have always functioned in dialogue with each other but in the future they will coalesce – most often in screen space where the possibility of receiving photomedia is increased – as is the capability of contributing to and changing the received photomedia imagery through these screens and the networks with which they are connected.

Aspects of these digital advances are currently in place but by 2039 they will have deepened and progressed further within the associated light-space-time structures of the epoch. Image-spaces that reside in inner-space will communicate and alter external image-spaces, this constant dialogue and potentially rapid change of all image-spaces will mean that these already shifting structures will transform faster through this constant exchange. This will have an impact on how we experience and live life in that understanding and perception will be constantly re-negotiated via new visual information conveyed on these screen spaces.

Overall, a substantial growth in computing power will allow high resolution photomedia images.[38] Also, a further extension of high processing power computers will be an increase in the possibilities for the manipulation of images and the generation of purely computer made images similar to those currently produced but with greater speed and flexibility.

Deep connections within an immense 'cloud' of computational resources means photomedia will, 30 years from now, be exceptionally powerful. Research scientist Alfred Spector claims that 'computer systems will have the opportunity to learn from the collective behaviour of billions of humans.'[39] He goes on to state that this 'intelligent cloud' will consist of millions of computers working to service the needs of billions of people in a constant feedback loop of queries and solutions. This same system could be used in the addition and recall of photomedia images.

The start of this direction of growth can be seen today but in the future these technologies will be greatly refined, forming part of common experience.

However, anticipating a future that is rich with sophisticated technological apparatus for the capture, storage and distribution of photomedia images is problematic. At the very least it signals one's own limitations of understanding and at best it predetermines fields of investigation for developing future systems. Whatever comes to be in 2039, it is certain that artists will be using it and responding to it.

The artist of 2039

New possibilities for the critique of experience, especially photomedia related experience, will evolve in this future. Daily experience will connect frequently with image-spaces and it is the artist's understanding of this that will be at the centre of many creative works. Like today, one may assume a scenario where specific concerns and methods allow artists to explore new interests or return to historically pertinent points in earlier artistic genres.

Also, like today, not all people in 2039 will embrace these technologies. For instance, in the late nineteenth century a movement in Germany which came to be known as the 'Lebensreform' (life-reform), embraced nature and a folk-spirit, rejecting the industrialisation and technology of the cities. These attitudes spread through the next century and found a major resurgence in the 1960s, in the 'hippy' movement which, much like the 'Lebensreform', sought to embrace 'natural' and older ways of living. And again in the 1990s, as digital technology emerged, a shift to a life closer to nature emerged in Australia, USA, UK and Canada in a group known as 'ferals'.[40] Essentially what these examples point to, and there are many others, is that when new technologies emerge some people reject them and return to older ways of living. This response is likely to continue into the 2030s. Some artists of 2039 will employ approaches that include a combination of 'high' and 'low' technology or 'obsolete' technology. Different photomedia technologies have specific characteristics, such as the texture and colour of particular films or papers and printing techniques, and artists will be attracted to these unique qualities. Attempts may also be made to mimic the specific qualities of older technologies through digital replications, but whatever occurs artists will use varying aspects of technology as a strategy to explore issues that may include but are not limited to, failure, redundancy, progress, experience and function – all issues that relate to technological advancement. It seems obvious that in a society which is experiencing exponential technological growth combined with ubiquitous image-spaces, these issues and a host of others will be engaged with by artists. Overall, as Jonathan Crary points out, 'we will need to consider the possibility of the coexistence of older and newer forms of vision, and above all, what this will mean for our ability to know, feel, and make sense of the world.'[41] Indeed, in the future there will be an increasing number of different points of view or interpretations of 'reality'.

Whatever the case, 'the history of the intersection of art and technology is one of the prognostications of an irrefutable, inevitable, and even immanent future that never comes to pass

– at least not exactly as we thought it might.'[42] Indeed, for all the affirmation given to the future of photomedia such as that which I have just put forward, when the time actualises it will no doubt be different in several if not many ways that cannot be known until 2039. From this point of view, the claim made by Sofia, Ono and Benjamin – that to imagine the future is to create it – is, at least, mediated in that when art and technology mesh they create previously unimaginable 'realities'.

Connected to nothing

The instantaneous capture and transmission of images is a dominant feature of the digital era; in 2039 the experience of these images in 'real-time' will persist. Future society will be perpetually connected to photomedia images and the associated image-spaces. In *Screened Out*[43] Baudrillard postulates that through the instantaneous transmission and reception of photomedia in 'real-time' we engage only with the image surface. Because of the speed with which the image is transmitted and the constancy with which new photomedia information is received, little time is left to move further than the surface of the image and to analysis and understand it completely. For Baudrillard this surface is an entry point, an opening into an empty space; a void which has the potential to be filled with anything.[44] Here lies the possibility of all images (the signifier), they can potentially depict anything as they are no longer confined to the material object (the signified); they are fantasies, manipulated 'realities', or totally computer constructed artefacts without any necessary connection to physical experience.

According to Baudrillard we have already left the obliteration of the real in our wake, advancing toward new modalities of experience that are no longer representations of the object, but simulations of the object.[45] If Baudrillard's vision is projected into the future, then this will no doubt be common experience. In addition, Baudrillard considers other modalities that are applicable to this future; they include the duplication of all objects which results in the destruction of the real by 'folding in on itself to the point of exhaustion'.[46] Baudrillard argues that this involves a process by which the real, the material 'object', is duplicated in serial form to such an extent that the real is no longer reflected in the image but eclipsed by it;

through serial production, multiples of the same are created which neither reflect nor refract their origin.[47] 'At the end of this process of reproducibility, the real is not only that which can be reproduced, but *that which is always already reproduced*: the hyperreal.'[48] This sees images function within themselves and of themselves, images of images, a system of representational models that produces what Baudrillard considers a state of dreaming, a state of hyperrealism. In the future, after continuing along this trajectory for some time, we will perhaps reach a point so subsumed by photomedia that we will be deep within image-worlds, a void space where meaning is of one's own making through the various relative connections and relationships between images.

Furthermore, a 'real-time' connectivity with everything signals the opposite – a connection with nothing; a relative void space of past and future time collapsing into to a perpetual now. In this future we are located in a digital space, what Baudrillard refers to as a 'luminous field of code'.[49] A digital realm of photomedia experience, a place of perpetual flux and gravitations, where images and their meaning and importance shift continuously.[50] If this is the case, we will find ourselves immersed within inner and outer-space, an image-space, without the capacity to distinguish between either space. It will be all around and within us, simultaneously forming and informing understanding. We will be floating in a void, separated from material reality.

In the 'photographic universe'

Flusser also sees us lost in a void, which takes the form of a 'photographic universe'. Even though Flusser formed his ideas concerning the photographic image in the 1980s he was a visionary who insightfully projected the progression of digital technologies, thus his arguments are relevant now and in the future. In this future of 2039, photomedia will be assimilated into everyday activities; its ubiquity will be visible and necessary due to our constant connection with it. As Flusser claims, this is a future 'turned into a global image scenario'.[51] Due to the constant interaction with the 'photographic universe' our lives will become a function of the images we create and a 'hallucinatory experience of the world' will result.[52] Unlike Flusser, Baudrillard critically examined the saturation of images first hand, from the 1980s until his death in 2007. He sees us engaging with a simulated reality that does not connect to reality itself, while Flusser does not even see an illusion of reality but a purely abstracted experience of that which does exist: the apparition of the 'photographic universe' and one's experience of it. While Baudrillard claims void spaces neither reflect nor refract their origin, Flusser claims that we are simply entrenched in the act of serving the apparatus that propagates a 'photographic universe', which compels us, in turn, to proliferate more images at the bidding of the technology. We will come to witness the overwhelming results of a society which is 'in a feedback relationship with the camera'.[53]

This process of the consumption of images and the potential contribution of one's own images produces image-worlds, a void-like space that is not empty but full of possibilities. Flusser speculated that we would be bound up in a cycle of 'feeding apparatuses and being fed by them'.[54] In 2039, we will have reached a point which allows for an information-based perspective that is photomedia based and where intrinsic value is subsumed by ascribed value. No longer will value be attributed to the content of the message, instead it will be attributed to the technology which transmits this message.

Flusser predicts that not only will society be working within the programmed 'freedom' of the apparatus, it will rely heavily on the limited program the machine initiates. Flusser states:

> Apparatuses are black boxes that simulate thinking in the sense of a combinatory game using number-like symbols; at the same time, they mechanize this thinking in such a way that, in future, human beings will become less and less competent to deal with it and have to rely more and more on apparatuses.[55]

As Flusser warns, the consequence of working in accordance with the program dictated by apparatuses is that society will be less able to work against them in the future. Since the inception of the photomedia image and the continual drive for technological improvement of the devices for making them, our relationships to the image would indicate that we are already in such a cycle.

Flusser theorises that society is so entrenched in this 'photographic universe' that as a whole we may find it difficult to move beyond the constraints of the programme. But, if we are to gain control of the 'programme' Flusser claims that first we need to understand that 'the fact that humankind is being programmed by surfaces (images) should not be considered a revolutionary piece of news.'[56] Indeed, reading a text is in some sense interacting with a 'surface'. The difference is, however, a surface text is 'static' whereas the progression of images is fluid. While we have been conforming to the program for some time, a recognition of this and a move away from it would be revolutionary. 'Photographs are programmed to model the future behaviours of their addresses. Yet, they are not only models of behaviour, but also models of perception and experience.'[57] Like Baudrillard, Flusser sees these surfaces as the location for information that is controlling our collective understanding. Like Baudrillard, Flusser sees us operating on the surface of these images; never going deeper and never understanding them. However, Flusser's surfaces are more like mosaics – highly fragmented and functioning in relation to each other in terms of temporal and spatial position. The opportunity to alter and play against the apparatus comes through creating the unexpected; this can occur on the most basic level by arranging these fragmented surfaces so that the juxtaposition of these fragments produces the illogical, similar to contemporary collage practice. By re-arranging the fragments that compose the seemingly seamless stream of images, the 'photographic universe' loses its hallucinatory coherence.

Pervading all understanding will be the photomedia image-space. These image-spaces will be an immersive aspect of our lives that will directly relate to experience, understanding and consciousness, as such they will shift and mutate due to our engagement with and contribution to them. It is a world composed from the geometry of fragments, less like mosaics and more like collages. Martha Rosler states:

> I am interested in all kinds of collages, but the kinds that interest me the most draw attention to spatiality, to the spatial dimension. That can mean an improbable relation of the fragments not only to each other but possibly also to the space within the frame, creating a "no space" or a contradictory one.[58]

In this way 'space' may well be re-created as a material reality as the viewer collides with the resistance of the 'illogical'. The future is the era of the supremacy of light, and light-spaces form image-worlds. These image-worlds formed from light-spaces will reconfigure so that, like collages, a 'no space' or unlikely light-space scenarios will form. In 2039 we will be in these spaces; the luminous fragments of photomedia will form and inform and be understood.

Despite Baudrillard's, Flusser's and Virilio's claims, I argue there is the possibility to see beyond the surface of the image. Throughout the history of photographic images, the eye and mind regularly go beyond these surfaces, allowing new ways of seeing that emerge from the connection between outer and inner-space, such as seen in Eadweard Muybridge's *The Horse in Motion* (discussed in Chapter 2) or Snow's *Wavelength* (discussed in Chapter 3). Many contemporary artists mentioned throughout this book such as Bill Viola, Gillian Wearing, David Noonan and Andro Wekua provide examples of work which allow the viewer to access a photographic space which is located within and also beyond the image, providing sign posts for the future of image-space. This is a process which moves beyond optics; it is still light-based but is lodged within psychological perception. That which moves between the synapses – these sparks of light – empower insight so that when the eye later encounters new images it can see these within the context of a new psychological perspective. This is the nature of the internal and external mechanisms of image-space. Runa Islam's video work CINEMATOGRAPHY (2007) is an example of this. Islam mounted a camera on a rig and through digital signals controlled the rig to make movements which gestured the letters C I N E M A T O G R A P H Y. The camera captured whatever was before the lens during these movements. The video work which is presented at the end of this process does not contain any of the techniques a cinematographer may employ such as depth of field, framing or lighting. Instead it shows only what occurred as the camera endured the mechanical gestures of the word *cinematography*. This approach by Islam reveals the materiality of the medium of film, and the nature of filmic representation, by allowing both of these aspects to be primary features of the work.

The future of the digital era will, like today, allow artists to intervene in the program by creating new processes and methods for analysis of photomedia. The only way to function beyond the system is to play against it, question it, knowingly expose the 'program' or set it aside. As Flusser asserts, 'Freedom is playing against the camera.'[59] To do this one must work beyond the programmed apparatus and produce the unexpected.

However, Flusser does not hold much hope for breaking away from this cycle in the digital future of 2039. He argues that it took centuries to understand that writing means storytelling and that it will take just as long to comprehend the systems of technological images. He claims, 'Our ignorance about the new codes is not surprising'.[60] Nonetheless, Flusser does provide a roadmap for understanding the situation better, claiming that the only path to liberation is a paradigm shift. This escape is only gained by understanding the language of technological images and using it to usurp the power of the apparatus. He explains that a photograph is not an image of the physical world as is generally thought, rather it is a combination of ideas; it is a non-linear world of concepts. Flusser asserts that the real revolution of photomedia is that they are 'models', and that an image contains within it concepts of the scene it depicts;[61] it is not the scene itself. Flusser warns that if we fail to decode photomedia 'we are condemned to endure a meaningless existence in a techno-imagery codified world that has become meaningless.'[62] The benefit of understanding photomedia is that we will no longer function unknowingly within a 'photographic universe'; instead, we will understand image-worlds, learn their 'language' and thereby gain control.

When speaking of the future, Flusser, in 1989, pointed out that, 'no more will we have to follow these models blindly. Instead, we will be actively engaged in the production of these models. It will be a universe through which we will project ourselves out of the present and into the future.'[63]

Flusser offers three steps toward this liberation: the first, to use our intelligence to outsmart the camera's rigidity, the second, to introduce human intentions into the program, and the third, to force the camera to produce the improbable and the unexpected. In this future those who have come to understand a philosophy of photography as Flusser would like us to, will work through a new paradigmatic process in an attempt to program the apparatus and society with a new vision. For as Flusser points out, the society of the future 'will be split into two classes: those programming and those being programmed. Into a class of those who produce programs and a class of those who behave according to the programs. Into a class of players and a class of puppets.'[64] If it is the case, that in the future viewer participation in the production of the image becomes commonplace (as discussed earlier in this chapter), then it is possible that society will be liberated and will produce authentic images free from the program of the apparatus.

At the speed of light

For Virilio the teleological end point is the speed of light. In 2039 the image will be received by the viewer faster and with greater frequency, increasing connections to other non-real spaces, events and communities and potentially engaging the global community in an unprecedented time Virilio calls 'global time'.[65] This 'global time' moves society away from localised time and into a generic 'time' which unfolds in a single, universal time. The result is not just a fundamental shift in a common understanding of any event, it is another profound shift in space and time. For Virilio, space has been superseded by time at the speed of light; the time it takes the photomedia image to be carried through space irrespective of real spatial distance. As technological devices for transporting imagery become more sophisticated, society will see a new world order, one based primarily on the photomedia image. Virilio suggests that we reach 'a general system of illumination that will allow everything to be seen and known, at every moment and in every place'.[66] And he argues, as does Flusser, in terms of 'players' and 'puppets', that power structures are certain to use this to their advantage.

Virilio's philosophy of 'dromology' becomes important when considering the structure of power within society for, as Virilio postulates, whoever controls a space possesses it and with this comes the power to control movement and circulation. Virilio contests that we will find ourselves within 'this dromocratic geocentrism of which we are far more the witnesses than the beneficiaries'.[67] Consistent with Flusser, Virilio suggests that those who control a space (physical or non-physical) gain power over those who do not. This becomes particularly relevant in online non-physical space, where laws cannot be written with enough speed to keep pace with changes as they occur. This scenario directs society toward a future where those who maintain and control non-physical spaces will have power equal to or greater than those who control physical space. The development of digital technologies has allowed an instantaneous transmission of imagery to the global population. Remote locations and events are experienced both 'here' and 'there' simultaneously and image content is increasingly contemporaneous. The technological aspects of this future which allows the photomedia image to gain greater omnipotence and power are significant. Anyone who has access (potentially everyone) may have influence by adding to the torrent of photomedia images. However, those who control access to the virtual space have the power to control those who engage with these spaces. Virilio does not see this as a democracy but rather a system of dominance.

Virilio warns that through the shortening of geophysical distance – the collapse of a distinction between psychological images (those images that exist in the inner-space of the mind) and photomedia images, and the multiple variations of this reality – a war of images will arise. This induces an epoch where 'everything passes through the image. The image has priority over the thing, the object, and sometimes the physically present being. Just as in real time, instantaneity has priority over space. Therefore the image is invasive and ubiquitous.'[68] Virilio highlights that photomedia engages society in a mediated experience whereby the image is supreme and far reaching, engaging in both physical and non-physical spaces to the point that subjective experience is an 'image' of an experience and not a material engagement.

The future, for Virilio, is part of a longstanding transport revolution, where the acceleration of the transmission of images will result in an 'accident'[69] and the loss of the material world. Virilio claims that in its own advancement the world has built in its own loss.[70] It has 'dematerialised' at the speed of light. We will be locked in a scenario of continuous and evolving accelerated and decelerated transmission of images, where speed becomes relative to prior speeds. Virilio claims that the result of this 'dromoscopy' is that, 'for the driver-prospector of the trip, the driver's seat is a *seat of prevision,* a control tower of the future of the trajectory.'[71] Throughout Virilio's theory of 'dromoscopy' the metaphor of the driver is often employed in reference to those who control the procession of images. Here Virilio considers the driver in terms of their capacity to explore, anticipate and ultimately control what is to come from the vantage point (prevision, or seeing the event before it is present) of the driver's seat.

If this is the case, then in 2039 we will be left in a catatonic state where 'Enlightenment fades away, along with the idea of the real, in the era of the speed of light.'[72] We are left on a journey through an endless array of images that takes the form of a hallucination.[73] These image-worlds will come to us like visions and without any connection to the real so that all we are left with is the acceleration and deceleration between images or the 'movement of movement'.[74] This not only effects the way images are transmitted, but also the meaning of the still image and the moving image.

As the technology continues to expand and mutate temporal and spatial possibilities, so it effects the still and moving image and the future viewer who alternates between streaming video and still photography. Peter Howe states, 'experience tells us that the still image not only remains in the viewer's subconscious but also engraves itself upon the psyche of the culture.'[75] Howe's example of such images includes Eddie Adams' photograph of the execution of a Vietcong suspect at the hand of a Vietnam police colonel. This unique function of the still image – to contain and memorialise the moment in a succinct manner – means that, despite the increased utilisation of the moving image, the still image may well retain a unique importance. The still image could even develop further conceptual weight as data streams gather speed. With the acceleration of information there will be a relative slowing of image mediated experience. I refer to Einstein's special theory of relativity which proposes that if one could travel at the speed of light, space would stretch out and time would move so slowly that it would seem still. As we reach a point in 2039 where the photomedia image is transported anywhere at the speed of light and streaming video imagery is everywhere, we will experience a relative slowing of light-time, which, while appearing slow, will in fact be moving very fast. In the future we will explore light-time with great speed but it will at times seem relatively slow but dense with images. As Nicolas Bourriaud claims, history is over, the past has collapsed into the present and 'the last continent to be discovered is time.'[76]

Carl Aigner suggests that 'the boundaries of visual images are the new boundaries of the world, and thus they are also the boundaries of life (and of all photography).'[77] What we make of these fragments that form a multifaceted and hallucinatory image world is not yet clear. Perhaps in an attempt to comprehend these future image-worlds we should start at the edges and move inward, that is start at the point where our perception of image-worlds begins so that we might ascertain what these spaces are. For instance, collage, through its abrupt collision of images arrests the smooth scanning of the image at the boundary of each fragment within the collage. One becomes aware of the 'images' themselves. It is from these metaphoric edges that we may begin to apply interpretations to their meaning and the intent behind them. By understanding images and their relationships to other images an understanding may form. This understanding begins through a knowledge that each image is composed of light and that this light carries coded information to be read, interpreted, understood and also to be challenged and altered. Perhaps in this way the 'viewer', that is all society, rises to Flusser's challenge to break free of the program.

This will be a world where the photomedia image will have intensified to such a degree that we will be simultaneously receiving and responsible for the mutation of these structures of light-space-time. Florian Rötzer believes, much like Virilio, that 'these images' increasing interactivity may soon supersede transportation in space, particularly since the time of spatial travel is annihilated by the transmission of images at the speed of light.'[78] We will have reached a stage in the evolution of photomedia where we no longer need to move in physical space to go somewhere, a point in the evolution of images where light at full speed takes us wherever we wish to go, as long as it is within the realms of non-physical space. Where we travel to and how we get there is, in part, up to us.

The future is here

Perhaps the most fascinating feature of this post-historical era is that, fundamentally, the intrinsic unifying element of photomedia technology will not change. Light will continue to be the essential agent in all photomedia; it will form our understanding and our thinking; it will shape our being. 'Light is the ultimate', writes Thomas Brill, 'its properties have influenced and, in part, controlled existence.'[79] In the future, light will dominate in the form of photomedia images; in the future of photomedia, light is everywhere, always. Society cannot escape it; we will find ourselves internally and externally bathed in Niépce's 'luminous fluid' as it formats light-space-time and in doing so affects experience and understanding. And so, this 200 year old story of the 'capture' of light will continue beyond 2039. However, if we wish to consider the future, the best place to start looking at it is in the present. As William Gibson claims, 'The future is already here, it's just unevenly distributed.'[80]

CONCLUSION

This book argues through a discussion of the history and future of photomedia, that light is not only a fundamental property of photomedia but that it binds with space and time to form and inform new experiential and perceptual structures. The effect of photomedia is so pervasive that it ultimately comes to shape understanding through image-worlds.

Despite the large body of theoretical enquiry into various aspects of photomedia, the importance of light to photomedia has been largely ignored.[1] This book forms part of a new, slowly growing awareness of this aspect of the media. It positions the reader in a new history and future which although overlooked by the canon of photomedia theory, is an essential line of enquiry for contemporary dialogue not least because of its future importance. While this book acknowledges the techno-historical aspects of photomedia, it is not tied to it; that half of this book examines contemporary and future conditions of photomedia highlights the relevance and importance of photomedia for common understanding.

In Chapter 1, an assessment of the period directly before the official invention of photography and the decades that immediately succeeded its invention – a period spanning from c.1830 to c.1880 – provided ground for a discussion of the nascent photomedia technology and its ensuing light-space-time structures. During this era, light attained a palpable presence; experience with light was discussed in terms of shaping visual perception and thus orientation in space and time. The technological, sociological and philosophical trajectories of Western society during this time were examined as a means of understanding the elements required for the invention of photography. Despite existing knowledge of the processes necessary for the invention, it was not until the industrial revolution that image machines came about. However, photomedia was not a machine which arose as a by-product of the industrial revolution. Rather, photomedia was the culmination of the efforts of the chemists, opticians and philosophers who understood and engaged with light in the eighteenth and nineteenth centuries. Early practitioners of light included Wedgwood, Talbot, Daguerre and Niépce. Wedgwood understood the power of light-based imagery to inform understanding and learning. Daguerre aimed to 'seize the light' and through his partnership with Niépce, who acquired a unique understanding of the transformative properties of light, developed the Daguerreotype.

Daguerre's image the *Boulevard du Temple, Paris* (c.1838) demonstrates a photographic time-shift, where the experiential space of the image displays a compressed light-time: a shift from the usual experience of space and time into a photographic time where past comes into the present. In this particular instance we see an elongated past of many moments made apparent through what is static, such as the man who stops to have his shoes shined, and what is dynamic and fleeting and unable to be captured, such as the busy Parisians traversing the Boulevard. A similar shift is demonstrated with Niépce's *Point of View* (c.1827) which is demonstrative of the way the photographic image functions in a temporally specific manner. It is representative of the new time which emerged in this era, photographic time. Both of these images demonstrate that photography, from its earliest expressions, binds light with space and time.

This capture of light in service of space and time is the beginning of what Virilio terms the industrialisation of vision. His dromoscopic model informs an understanding of these early images and their inherent displays of light-space-time. Virilio theorises a logic of speed and its impact.

Another important early manifestation is discussed: the ability for the photograph to frame a particular View...such as seen in Talbot's *Latticed Window*. Talbot treasured realism and saw the photograph as a transcription of the real. But, as Baudrillard's theory of simulacra claims, due to the proliferation of the photographic image a disappearance of reality occurs.

The manifestation of a new expression of light-space forms image-spaces and gives rise to image-worlds which arise from the dissemination of photographic images; a cyclical mechanism of image production which informs reception and in turn forms production: inner personal space meshes with outer public space as seen in Bayard's *Self Portrait as a Drowned Man*. This dynamic connects with Flusser's theory of the 'photographic universe'. Flusser claims that cameras are used not only to capture the otherwise imperceptible realms of inner-space but also that which lies beyond perceptual boundaries in outer-space and in the micro realm such as that demonstrated by Dr Luys' *Four diameter cross-section of segments of the cerebellum* (c.1870). Dr Luys explored the microscopic realms of the brain with his images taken through a microscope, while the macro realms of the universe are explored by Whipple with his image *Moon*. Each image brings these otherwise imperceptible realms proximately close through optical technologies and photographic techniques. This cataloguing of physical reality results in a perceptual contraction of geophysical space which connects with Virilio's theory of 'dromology', a theory of speed. In this era photography demonstrates the first jolt of speed which Virilio sees as the beginning of a loss of space through the acceleration of time induced by the photographic image. However, this is an era of slow time, when luminous duration becomes impressed upon the photographic surface, bearing down with an intense physicality. Overall, this chapter explores the first light-space-time structures and how they manifest in the earliest images.

During the second technological era (c.1871–c.1990), the analogue era, discussed in Chapter 2, I argue that light-space-time structures configure into formations particular to the period. Through the increasing technologising of light, and the various emerging photomedia technologies, light-time accelerates the speed and transmission of light-spaces. Instantaneous photography which captured light at a fraction of a second emerged in this era. With this new development in photomedia Muybridge, through his studies of physical movement such as *The Horse in Motion* (1878), revealed space and time beyond what the eye could normally see. Marey achieved the effect of collapsing time into a moment in light-space with images such as *Movements in pole vaulting* (c.1900). Edgerton used instantaneous photography to represent time in discreet and quantised forms as seen in *Bullet through Banana* (1964). Marey and Edgerton each demonstrate the precise moments of light-time revealing what cannot normally be seen and thus changing one's understanding of time, irrevocably. The Photofuturist movement demonstrated the reanimation of light-time, seen in the work of Bragaglia, who explored the fluid, chaotic expression of movement through space and time in images such as *The typist* (1911). This image produced a sensation of fast-time as a representation of the speed of progress of the era. Another aspect is the light-space-time of cinema, as explored in the work of Maholy-Nagy, who considered photomedia a light-form. His light-space modulator allowed him to paint with light in the formation of his film *Lichtspiel Schwarz-Weiss-Grau* (1930), a dynamic visualisation of light-forms through space and time. The introduction of cinema, as claimed by David Campany, brought about a new dichotomy between movement and stillness.

In this era, the still image gained greater weight in memorialising the moment, allowing the viewer to reflect on the intricacies of light-space held in the fast capture of light-time. Snow's *Wavelength* (1967) explored the limitations and boundaries of movement and stillness. Marker's *La Jetée* (1962) highlights the power of the still image to convey inner and outer image-space; the still becomes the moving. Whereas in Andy Warhol's *Empire* (1964) the moving is explored as the still. Sherman responds with her own understanding as seen in *Untitled Film Still* (1978), while Sugimoto's images of movie theatres such as *Ohio Theatre, Ohio* (1980) allow us to think of the vectors of these image-spaces, such as speed, light and time through space. In Flusser's analysis speed separates space from time and in Virilio's philosophy of dromololgy time comes to dominate space. Ultimately, with the fast transfer of images, the image-spaces of the analogue era establish the ubiquity of the photomedia image.

Flusser warns against engaging with the photographic act, and encourages society to decode it as a language. He argues that all images circulate within society as an indicator of image-space, mass culture and stereotypical experiential reality, an image-world. As Flusser claims, the apparatus of photography will continually improve itself and in doing so will continue to control society to initiate this improvement. This only continues to proliferate an image world which both immortalises the photographer and continually advances photomedia. Virilio claims that through photographic images, particularly those that aim to transcribe the world, we have lost touch with the real. Each theorist sees the proliferation of the photomedia image as a method for obscuring our connection with the world.

Cartier-Bresson is an example of a photographer who believes in the photograph as a signifier of events, best understood in terms of his idea of a 'decisive moment' and which can be seen in *Place de l'Europe, Paris* (1932). This photograph conveys light-space-time structures and its place in the world contributes to the always shifting image-world. Many possible combinations of interpretations may arise from engagement with this image-world. What is clear in this era is that there is no single perspective or concrete example of what is real. Collage presents a new strategy in art making, allowing engagement with the fragmented image-spaces of the era as much as challenging them. This is demonstrated in Heartfield's *Hurrah, die Butter ist alle!* (1935) and in Rosler's *Red Stripe Kitchen* (1967–1972).

Video art further engaged with light-space-time and was used by artists such as Paik who, in *Moon is the oldest TV* (1965), explored the illusory qualities of the medium to represent physical reality. Warhol, who in *Outer and inner space* (1966), explored image-space and the inner and outer experience that define the encounter with photomedia. Buky Schwartz engages further with perceptual space by bringing together physical and video space in *Yellow Triangle* (1979). I end the chapter by asking what occurs when the material retreats towards the immaterial and seek to answer this question in part through consideration of Yves Klein's *Leap into the void* (1960). However, this question gains particular pertinence in the next era and is therefore more fully addressed in the next chapter.

In Chapter 3, the digital era (c.1990–2013) is discussed, establishing that, despite an anxiety about the death of photography at the introduction of digital photomedia, the technology did not fundamentally alter the light-based nature of the medium. However, the digital era erases the possibility of an indexical connection between the image and that which it represents. Baudrillard's understanding of the void through the absent connection between the sign contained within the photomedia image and the thing it represents is analysed, as is Virilio's theory of speed and transition to the void through photomedia and its fast transmission. Flusser sees the void as a process of stepping away from reality through the 'photographic universe'. However, the digital is not a post-photographic era, but a continuation and development of the light-space-time structures that preceded it. This chapter illustrates that in the digital era light-space-time developed with increasing vigour as image-spaces gained convincing and malleable pictorial compositions and highly compelling constructs via computer technology. Digital photomedia extends the possibility of constructed scenarios through the manipulation of the photograph as light-based information. No longer is something required to physically exist to be presented as such. The result is improved malleability of the image, as seen in Jeff Wall's *After 'Invisible Man' by Ralph Ellison, the Preface* (1999–2000), Crewdson's *Untitled 'Beneath the Roses'* (2004) and Charlie White's *Champion* (2005).

Photomedia, which was progressively found on luminous screens, formed a new light-space. This screen space now combines with the computing power of digital photomedia to proliferate light-space-time structures quickly and responsively. Inner- and outer-spaces have come together in photomedia image-spaces, and are displayed on luminous screens with increasing frequency, a phenomenon closely connected with the philosophy of Huhtamo. Specifically, this has been a relationship that has long existed before digital screens but I theorise that now we carry with us the fragments of our encounters with screens that both reflect and refract light. Light-time gained instantaneity, as Beaubois explores in his successful connection of speed with vision with *The fall from Raiatea* (2007).

As the technological drive for faster ways to take and present the image evolved, a new light-time experience emerged: slowness because of speed. Artists engaged with this paradoxical slowness, such as Wearing who responded with *Snapshot* (2005) which explores the stillness of the photograph through moving video. Bill Viola responded with *The fall into paradise* (2005) which uses extremely slow motion to enter time itself and engage with the expressive moments normally lost in real time, inducing the sensation of the passage of time and the constant desire to understand each moment while it slips away. This work also connects with Yves Klein's *Leap into the void* (1960) and the intangible forces that come to affect us increasingly more than the materially tangible. In the digital epoch greater fragmentation was demonstrated as the perpetually shifting image-spaces are transmitted and received through the light-time of instantaneous transmission. These image-spaces moved closer, faster as this digital era embraces a new geometry of light-space-time. Virilio sees this fragmentation as one that occurs in inner as well as outer-space. Collage, which offers a strategy for freedom from the programming apparatus that Flusser theorises, is employed in the digital era by artists such as David Noonan with *Kabarett Keif* (2007) and Andro Wekua with *Lying, Walking, Swimming* (2005). Collage allows the use of images in a new unexpected way, as explored by Daniel Crooks with *Train No. 1* (2002–2005) with his collage like cuts into the video image. Ultimately the digital era is characterised by the instantaneous dissemination of light-time and its antithetical expression, slowness and extensive fragmentation of image-space.

In Chapter 4, I briefly discuss the associated problems in predicting the future. Any analysis of the future is determined by one's understanding of their current situation and their utopian or dystopian perspectives for the future. I next examine the potential technological growth of photomedia and the direction this may take. The year 2039, 200 years after the official invention of photography, is established as a notional future benchmark for speculating how photomedia might develop. Changes in relation to the light-space-time structures move and form with agility and speeds never before seen. This future sees light-time reach an unprecedented speed which incurs a relative slowing of the experience of light-time. Also this epoch sees a complex image-space developing where geophysical space and real time will fall away as light-space-time dominates. Future experience and understanding influenced by these image-spaces, and the fragmented and collaged light-space which ensues, will bring us to a no-place and a no-space of no-time. This is a void, an image-space created through engagement and participation with the image-world. This will be an era of the primacy of light, a matrix for the creation of new experiences and realities which will initiate a multifaceted and hallucinatory image-world. The future era, while initiating new paradigms, will not end the 200 year old story of photomedia; rather it will be a further development of the technology and our current experiences.

Baudrillard sees the future in terms of simulated reality, which becomes a void of nothing and zero connection to the real. Flusser and Virilio also see us passing into a void. Flusser sees us lost in a 'photographic universe' and Virilio understands the future in terms of the speed of light and 'dromoscopy'. Essentially this future is already with us in many ways and although it may have points of difference, it will most certainly have points of similarity. Light has always been the defining characteristic of photomedia and this will continue in the future.

By setting out this light-based theory of photomedia, new possibilities and questions will hopefully arise. With these questions, unexpected responses and new questions may emerge. This is important because through this book fresh dialogue concerning photomedia and experience in general may be generated. With the increasing proliferation of photomedia within contemporary art and in predicted future experience it is important to continually question the assumptions one may have concerning photomedia.

Notes

Introduction

1. Other research on this area includes Sean Cubitt, who discusses light in terms of its technological application, See: Sean Cubitt, 'New Light', in *Transforming Aesthetics*, ed. Jill Bennett and Anna Munster (University Presses of New England, 2007). Also, Melissa Miles who examines analogue photography and its relationship to light, see: Melissa Miles, *The Burning Mirror: Photography in an Ambivalent Light* (North Melbourne, Vic.: Australian Scholarly Publishing, 2008). Also, Michel Frizot, 'Light Machines: On the Threshold of Invention', in *A New History of Photography*, ed. Michel Frizot (Cologne: Könemann, 1998).
2. See: Ian McNeil, *An Encyclopaedia of the History of Technology* (London; New York: Routledge, 1990). Also, Beaumont Newhall, *The History of Photography: From 1839 to the Present*, Completely rev. and enl. ed. (London: Secker & Warburg, 1982). Also, Mary Warner Marien, *Photography: A Cultural History* (London: Laurence King, 2002), which includes an appendix detailing a chronology of photographs and major artistic, cultural and political developments from 1510 to 2002.
3. Such as Kodak: Todd Gustavson, *Camera: A History of Photography from Daguerreotype to Digital* (New York: Sterling Publishing, 2009). Also, http://wwwau.kodak.com/global/en/corp/historyOfKodak/historyIntro.jhtml?pq-path=2217/2687; Fuji, see: http://www.fujifilm.com.au/extrapages/default.asp?id=7; Canon, see: http://www.canon.com/about/history/; and Nikon, see: http://www.nikon.com/about/info/history/index.htm (all accessed 25 May 2010).

4. *Understanding Moore's Law: Four Decades of Innovation*, ed. David C. Brock (Philadelphia: Chemical Heritage Press, 2006).
5. French, 1929–2007. See Jean Baudrillard, *Screened Out*, trans. Chris Turner (London: Verso, 2002). Jean Baudrillard, *Photography, or the Writing of Light* (CTheory.net, 2000 [cited 6 June 2006]); available from http://www.ctheory.net/articles.aspx?id=126. Jean Baudrillard, 'Precession of Simulacra,' in *Simulacra and Simulation* (Ann Arbor: University of Michigan Press, 1994). Jean Baudrillard, *Simulacra and Simulation*, trans. Sheila Faria Glaser (Ann Arbor: University of Michigan Press, 1994). Jean Baudrillard, *Rise of the Void Towards the Periphery* (CTheory.net, 1994 [cited 6 June 2006]); available from www.ctheory.net/text_file?pick=58 (accessed 6 July 2007). Jean Baudrillard, *The Art of Disappearance*, trans. Nicholas Zurbrugg (Brisbane: Institute of Modern Art, 1994). Jean Baudrillard, *Symbolic Exchange and Death*, trans. Iain Hamilton Grant (London: Sage in association with Theory, Culture & Society, School of Health, Social and Policy Studies, University of Teesside, 1993). Jean Baudrillard, *Simulations*, trans. Paul Foss, Paul Patton and Philip Beitchman (New York: Semiotext(e), 1983).
6. Born in Czechoslovakia, lived in Brazil and France, 1920–1991. See: Vilém Flusser, *Towards a Philosophy of Photography*, trans. Anthony Mathews (London: Reaktion, 2000). Vilém Flusser, *The Shape of Things: A Philosophy of Design*, trans. Anthony Mathews (London: Reaktion, 1999), Vilém Flusser, *Vilém Flusser: Writings*, ed. Andreas Ströhl, trans. Erik Eisel (Minneapolis: University of Minnesota Press, 2002).
7. French, b.1932. See: Paul Virilio, *Art as Far as the Eye Can See*, trans. Julie Rose, English ed. (New York: Berg, 2007). Paul Virilio, *Negative Horizon: An Essay in Dromoscopy*, ed. Michael Degener, trans. Michael Degener (London; New York: Continuum, 2005). Paul Virilio, *Desert Screen: War at the Speed of Light*, trans. Michael Degener (New York: Continuum, 2002). Paul Virilio, *The Information Bomb*, trans. Chris Turner (London: Verso, 2000). Paul Virilio, 'Red Alert in Cyberspace,' *Radical Philosophy*, no. 74 (1995). Paul Virilio, *The Vision Machine* (Bloomington: Indiana University Press, 1994). Paul Virilio, *The Aesthetics of Disappearance*, trans. Philip Beitchman, 1st English ed. (New York: Semiotext(e), 1991).
8. A media and warfare theory, which proposes a science or logic of speed and a study of its impact. See: Virilio, *Negative Horizon*.
9. The 'photographic universe' encompasses all of the photographs thus produced. Flusser, *Towards a Philosophy of Photography*.

10. Chronophotograph was a word coined by Étienne-Jules Marey to describe his photographic studies of motion; the term is used in reference to photographs which depict chronological movement. At times such movement is segmented as in Muybridge's work and at other times the complete range of movement is shown over a durational exposure as in Marey's work.
11. David Campany, 'Safety in Numbness: Some Remarks on Problems of "Late Photography",' in *Where Is the Photograph?*, ed. David Green (Maidstone, Kent; Brighton: Photoworks; Photoforum, 2003).
12. Virilio, *Negative Horizon.*
13. Flusser, *Towards a Philosophy of Photography.*
14. Baudrillard, *Symbolic Exchange and Death.*
15. Paul Virilio and Sylvère Lotringe, *Pure War*, trans. Mark Polizotti (New York: Semiotext(e), 1983). Also, Virilio, *Negative Horizon.*
16. Flusser, *Towards a Philosophy of Photography.*
17. Ibid.
18. Baudrillard, *Screened Out.* Also, Baudrillard, *Rise of the Void* and, Baudrillard, *Symbolic Exchange and Death.*
19. Virilio, 'Red Alert in Cyberspace', *Desert Screen* and *Negative Horizon.*

Chapter 1

1. Joseph Nicéphore Niépce cited in Paul Virilio, *The Vision Machine* (Bloomington: Indiana University Press, 1994), p.20.
2. French, 1765–1833.
3. Vicki Bruce, *Visual Perception: Physiology, Psychology and Ecology*, ed. Patrick R. Green and Mark A. Georgeson, 4th ed. (Hove: Psychology, 2003).
4. For an overview of the process of visual perception see R. L. Gregory, *Eye and Brain: The Psychology of Seeing*, 4th ed. (Oxford; New York: Oxford University Press, 1994).
5. Ibid. p.17.
6. Latin for 'veiled chamber', a camera obscura is essentially a dark chamber with an aperture for passing light through. Once this light is reflected on a surface inside the chamber, a view of the external scene is shown upside down.
7. The history and legitimacy of the 'first' inventor of the camera obscura are varied. Some historians claim the inventor of the camera obscura to be Aristotle, others Roger Bacon, Leonardo da Vinci or Giambattista Della Porta among others. Della Porta was an important exponent of the camera obscura, popularising knowledge of the

device in his book *Magiae Naturalis; sive, De miraculis rerum naturalium* (1538) or as it was commonly known through the English translation of the text, *Natural Magick*. For more information on the history of the camera obscura see Josef Maria Eder, *History of Photography*, trans. Edward Epstean (New York: Dover Publications, 1978).

8. This has been suggested by many, including the following: Naomi Rosenblum, *A World History of Photography*, 1st ed. (New York: Abbeville Press, 1984), p.15; Peter Galassi, *Before Photography: Painting and the Invention of Photography*, ed. Museum of Modern Art, trans. Museum of Modern Art (New York; Boston: Museum of Modern Art; Distributed by New York Graphic Society, 1981), p.11; and Mike Ware, 'On Proto-Photography and the Shroud of Turin', *History of Photography* 21, no. 4 (1997), pp.261–9.
9. German, 1684–1744.
10. Jean-Claude Lemagny and André Rouillé, *A History of Photography: Social and Cultural Perspectives*, trans. Janet Lloyd, English language ed. (Cambridge; New York: Cambridge University Press, 1987), pp.12–3.
11. Richard Buckley Litchfield, *Tom Wedgwood, the First Photographer; an Account of His Life, His Discovery and His Friendship with Samuel Taylor Coleridge, Including the Letters of Coleridge to the Wedgwoods and an Examination of Accounts of Alleged Earlier Photographic Discoveries* (London: Duckworth, 1903), p.224.
12. Essentially, the method requires the photographic substrate to be coated in an emulsion of silver salt particles. The formation of the silver salts is unique for each image; this aspect, however, is usually only visible at a microscopic level. When light of a particular wavelength is absorbed by the emulsion, an excitation of the emulsion at an atomic level produces a silver salt grain. Light is absorbed in the image, changes the material in which it interacts and materialises as shades of light as seen in the spectrum from white through greys to black on the array of silver salts. Because the silver salts are irregularly shaped, their formation and that of the shades of light and dark are particular to each image, demonstrating that the image as a carrier of light is unique on this microscopic level.
13. Michel Frizot, 'Light Machines: On the Threshold of Invention', in *A New History of Photography*, ed. Michel Frizot (Cologne: Könemann, 1998).
14. English, 1792–1871.
15. Although light does not actually write (or draw) an image, the name 'photography' was used to describe the way light alters silver salts (nitrate

and chloride) and provides a representation of the world through the gradient shades of black and white. See Andrew E. Hershberger, 'The Past, Present and Future of the History of Photography: Interviews with Peter C. Bunnell, Gretchen Garner and Britt Salvesen', *History of Photography* 30, no. 3 (2006), p.204.

16. English, 1771–1805.
17. English, 1800–1877.
18. Geoffrey Batchen, *Burning with Desire: The Conception of Photography* (Cambridge, MA: MIT Press, 1997), p.35.
19. A letter by Talbot in Ann Thomas, *Beauty of Another Order: Photography in Science* (New Haven: Yale University Press in association with the National Gallery of Canada, Ottawa, 1997), p.29.
20. Ibid. p.33.
21. French, 1787–1851.
22. Herr Schiendle in Litchfield, *Tom Wedgwood, the First Photographer*, p.226.
23. Ibid. p.226.
24. Galassi, *Before Photography*, p.11.
25. For more information see ibid.
26. Thomas, *Beauty of Another Order*, p.29.
27. Galassi, *Before Photography*, p.11.
28. Batchen, *Burning with Desire.*
29. Ibid, p.161.
30. Herbert Ohlman, 'Information: Timekeeping, Computing, Telecommunications and Audiovisual Technologies', in *An Encyclopaedia of the History of Technology*, ed. Ian McNeil (London; New York: Routledge, 1990), p.691.
31. Lemagny and Rouillé, *A History of Photography.*
32. Lemagny and Rouillé discuss how art of the Renaissance was attempting to refine the representation of space and volume on a flat surface, which lead to the development of sophisticated methods of representing perspective and the 'faithful' representation of objects in space. See ibid, p.14.
33. Ibid, p.15.
34. Galassi, *Before Photography*, p.11.
35. Tim Cloudsley, 'Romanticism and the Industrial Revolution in Britain', *History of European ideas* 12, no. 5 (1990), p.611.
36. Laura Mulvey, *Death 24x a Second: Stillness and the Moving Image* (London: Reaktion Books, 2006), p.18.
37. The chemists, scientists and opticians that Thomas refers to include but are not limited to Humphry Davy (British 1778–1829), Louis-Jacques-Mandé

Daguerre and William Henry Fox Talbot. See Thomas, *Beauty of Another Order*, p.76.

38. Tom Wedgwood to William Godwin, 31 July 1797. Quoted by W. St. Clair, *The Godwins and the Shelleys, the Biography of a Family* (London: Faber and Faber, 1989).
39. Thomas Wedgwood is a fascinating character; regrettably, this book does not allow a complete exploration of his life and his network of friends and family as they too are remarkable. For more biographical information on Wedgwood two primary texts remain the most comprehensive to date, the first being Litchfield, *Tom Wedgwood, the First Photographer* and the second, Eliza Meteyard, *A Group of Englishmen (1795 to 1815): Being Records of the Younger Wedgwoods and Their Friends, Embracing the History of the Discovery of Photography and a Facsimile of the First Photograph* (London: Longmans, Green, and Co., 1871). A broader but equally fascinating read is B. and H. Wedgwood, *The Wedgwood Circle, 1730–1897, Four Generations of a Family and Their Friends* (Westfield, NJ: Eastview Editions, 1980).
40. The announcement was made in the Academy's official publication the *Cômpte-rendu des Séances de l'Académie des Sciences.* An English translation of the announcement was published in the 19 January edition of the *Literary Gazette.* For more information see Beaumont Newhall, *The History of Photography: From 1839 to the Present*, completely rev. and enl. ed. (London: Secker & Warburg, 1982), p.19.
41. Ibid.
42. Edgar Allan Poe points out in his essay 'The Daguerreotype' (1840), that the correct spelling is Daguérreotype, and 'pronounced as if written Dagairraioteep' in *Classic Essays on Photography*, ed. Alan Trachtenberg, trans. Alan Trachtenberg (New Haven, CT: Leete's Island Books, 1980).
43. Initially the invention was published in French, then in English a week later.
44. For more information about the Daguerreotype and its use in France and America respectively see: Janet E. Buerger, *French Daguerreotypes*, ed. International Museum of Photography at George Eastman House (Chicago: University of Chicago Press, 1989). Also, Beaumont Newhall, *The Daguerreotype in America*, 3rd ed. (New York: Dover Publications, 1976).
45. English, 1801–1885.
46. Daguerre cited in Melissa Miles, 'Focus on the Sun: The Demand for New Myths of Light in Contemporary Australian Photography', *Australian and New Zealand Journal of Art* 9, no. 1/2 (2008/2009), p.222.

47. Newhall, *The History of Photography*, p.15.
48. Daguerre's apprenticeship also included architecture and theatre design. Daguerre became well known for his skill in theatrical illusion, and while working as a set designer for theatre he developed the Diorama. For more information see: Helmut Gernsheim, *L. J. M. Daguerre: The History of the Diorama and the Daguerreotype* (London: Secker and Warburg, 1956).
49. For a thorough account of Monet's work and life see Christoph Heinrich, *Claude Monet, 1840–1926*, trans. Michael Hulse (Cologne: Benedikt Taschen, 2000).
50. Daguerre did not intentionally seek the capture of a human presence within what is otherwise a panoramic image of the Boulevard. The first intentional portrait was taken by Samuel Finley Breese Morse (American, 1791–1872). Morse, the inventor of the electric telegraph and Morse code, met Daguerre in Paris in 1838. On his return to America, Morse took up photography as an extension of his understanding of portrait painting and sculpture. In 1939, another photographer, John William Draper captured a portrait image by covering his sitter's face with flour to increase luminosity and exposed the image for 30 minutes. For more information see: Brian Clegg, *Light Years: The Extraordinary Story of Mankind's Fascination with Light*, ed. Lizzie Hutchins (London: Piatkus, 2001).
51. Virilio, *The Vision Machine*, p.19.
52. His metaphors are so evocative of the nature of light that physicists of the period used them for promoting their concerns regarding electricity and electromagnetism. Ibid, p.20.
53. The image shown in Figure 1.2 is an enhanced reproduction of the original. The original image is so delicate that it cannot be displayed in direct light, and on the rare occasions it is shown, the image must be viewed in a controlled lighting situation. However, this reproduction allows for analysis of the light-space-time qualities of the original.
54. Heliography when translated means 'sun writing'.
55. Virilio, *The Vision Machine*, p.20.
56. Ibid, p.19.
57. Paul Virilio, *Negative Horizon: An Essay in Dromoscopy*, ed. Michael Degener, trans. Michael Degener (London; New York: Continuum, 2005).
58. Ibid, p.105.
59. Paul Virilio, *The Virilio Reader* (Malden, MA: Blackwell Publishers, 1998), p.72.
60. Virilio, *Negative Horizon*, p.30.

61. *Art and Philosophy: Baudrillard, Gadamer, Jameson, Kristeva, Lyotard, Marin, Perniola, Sloterdijk, Sollers, Virilio, West*, ed. Catherine Francblin and Jean Baudrillard, trans. Catherine Francblin and Jean Baudrillard (Milan: G. Politi, 1991), p.145.
62. Ibid, p.149.
63. Thomas, *Beauty of Another Order*, p.76.
64. In *Essays on Remarkable Photographs* Sophie Howarth discusses Talbot's *Latticed Window* and highlights a correlation between Niépce's first photograph, Talbot's *Latticed Window* and the contemporary photograph by Jeff Wall, *A view from an Apartment 204–5*. Howarth states that they are all, 'a metaphor for the way in which photography literally frames a view onto the world...and in each case the image has been laboriously constructed by experimenting with the latest technological possibilities.' See Sophie Howarth, ed., *Essays on Remarkable Photographs* (New York: Aperture, 2005).
65. Annotation on the back of Talbot's image *Latticed Window*, shown in Figure 1.3.
66. See Jean Baudrillard, *Simulations*, trans. Paul Foss, Paul Patton and Philip Beitchman (New York: Semiotext(e), 1983). Also, Jean Baudrillard, *Symbolic Exchange and Death*, trans. Iain Hamilton Grant (London: Sage in association with Theory, Culture & Society, School of Health, Social and Policy Studies, University of Teesside, 1993); Jean Baudrillard, *Simulacra and Simulation*, trans. Sheila Faria Glaser (Ann Arbor: University of Michigan Press, 1994); Jean Baudrillard, 'Precession of Simulacra', in *Simulacra and Simulation* (Ann Arbor: University of Michigan Press, 1994).
67. Jean Baudrillard, *Photography, or the Writing of Light* (CTheory.net, 2000 [cited 6 June 2006]); available from http://www.ctheory.net/articles.aspx?id=126.
68. Plato as cited in Ibid.
69. Jean Baudrillard, *The Art of Disappearance*, trans. Nicholas Zurbrugg (Brisbane: Institute of Modern Art, 1994), p.2.
70. Batchen, *Burning with Desire*, pp.158–173.
71. Batchen cites that a coup was attempted earlier that year and Adolphe Thiers, the Prime Minister, was pressured into resigning three days after Bayard's image was made. See ibid, p.167.
72. Ibid, p.167.
73. Bayard had exhibited his photographic process before Daguerre; however he was duped into keeping the process a secret by François Arago (French, 1786–1853), a friend of Daguerre's. This gave Daguerre the necessary time to finalise his invention and present it

to the French Academie des Sciences with no competition. See Clegg, *Light Years.*

74. Gaston Bachelard, *The Poetics of Space*, trans. Maria Jolas (Boston: Beacon Press, 1969).
75. Baudrillard, *Photography, or the Writing of Light.*
76. Larry J. Schaaf, *Out of the Shadows: Herschel, Talbot, and the Invention of Photography* (New Haven and London: Yale University Press, 1992), p.45.
77. French, 1828–1895.
78. The cerebellum is a part of the brain which controls sensory perception, coordination and motor skills.
79. Dr Jules Bernard Luys referenced in Thomas, *Beauty of Another Order*, p.100.
80. English, 1792–1871.
81. See John F. W. Herschel, 'On the Action of the Rays of the Solar Spectrum on Vegetable Colours, and on Some New Photographic Processes on the Action of the Rays of the Solar Spectrum on Vegetable Colours, and on Some New Photographic Processes', *Philosophical Transactions of the Royal Society of London* 132 (1842), pp.181–214. Also, John F. W. Herschel, 'On the Chemical Action of the Rays of the Solar Spectrum on Preparations of Silver and Other Substances, Both Metallic and Non-Metallic, and on Some Photographic Processes on the Chemical Action of the Rays of the Solar Spectrum on Preparations of Silver and Other Substances, Both Metallic and Non-Metallic, and on Some Photographic Processes', *Philosophical Transactions of the Royal Society of London* 130 (1840), pp.1–59.
82. Thomas, *Beauty of Another Order*, p.150.
83. The American theorist on the Daguerreotype Court Justice Oliver Wendell Holmes (1809–1894) referred to the Daguerrotype as a 'mirror with a memory'. See Hershberger, 'The Past, Present and Future of the History of Photography, p.207.
84. Virilio, *Negative Horizon*, p.120.
85. For more information on 'visual retention', see Virilio, 'The vision machine', *The Virilio Reader*, pp.134–151.
86. Virilio, *The Vision Machine.*
87. André-Adolphe-Eugène Disdéri (French, 1819–1889) was a photographer who invented the cartes de visite, a small photographic calling card. See Kelley E. Wilder, 'Disdéri, André-Adolphe-Eugène', in *The Oxford Companion to the Photograph*, ed. Robin Lenman (Oxford: Oxford University Press, 2005).
88. Virilio, *The Vision Machine*, p.21.

89. The pseudonym for the Frenchman Gaspard-Félix Tournachon, (1820–1910).
90. Virilio, *The Vision Machine*, p.20.
91. Vilém Flusser, *Towards a Philosophy of Photography*, trans. Anthony Mathews (London: Reaktion, 2000).
92. Ibid.

Chapter 2

1. Analogue photomedia, by definition, records light as a chemical process or electric non-digital signal. I define the analogue age of photomedia to occur from 1870 to 1990. See *Van Nostrand's Scientific Encyclopedia*, ed. Douglas M. Considine, 7th ed. (New York: Van Nostrand Reinhold, 1989).
2. Anschütz also took photographs of physical movement, gaining a reputation for his images of military manoeuvres. See *Encyclopedia of Early Cinema*, ed. Richard Abel (London; New York: Routledge, 2005), pp. 29–30.
3. English, 1830–1904.
4. French, 1830–1904. Coincidentally Muybridge and Marey lived the same life span; they were born within a month of each other and died within a few weeks of each other, both aged 74 years.
5. French, 1858–1917.
6. English, 1852–1916.
7. American, 1903–1990.
8. Ann Thomas, *Beauty of Another Order: Photography in Science* (New Haven: Yale University Press in association with the National Gallery of Canada, Ottawa, 1997), p.150.
9. Chronophotograph was a word coined by Étienne-Jules Marey to describe his photographic studies of motion. The term is used in reference to photographs which depict chronological movement; at times such movement is segmented as in Muybridge's work and at other times the complete range of movement is shown over a durational exposure as in Marey's work.
10. For more information on the process Muybridge used see Michel Frizot, 'Speed of Photography: Movement and Duration', in *A New History of Photography*, ed. Michel Frizot (Cologne: Könemann, 1998), pp.243–57.
11. Thierry de Duve, 'Time Exposure and Snapshot: The Photograph as Paradox', *October* 5 Summer (1978), p.115.
12. The Zoopraxiscope was a proto-cinematic device invented by Muybridge in 1879. It animated movement like the Zeotrope, a common

parlour game of the time which used spinning disks of images which moved in one direction while the viewer looked at them them through another disk spinning in the opposite direction with four black slots. However, the Zoopraxiscope extended this process by projecting moving images outwards, like a modern movie projector. This allowed many people to see the same moving image simultaneously. Also, the Zoopraxiscope used rotating glass disks to show sequentially placed photographic images in fast succession. Ultimately the Zoopraxiscope directly influenced Thomas Edison to later invent the Kinetoscope, and develop the motion picture.

13. Mart Braun, 'Animal locomotion' in Marta Braun, Stephen Herbert, Paul Hill, Anne McCormack, Kingston Museum and Heritage Centre, *Eadweard Muybridge: The Kingston Museum Bequest*, ed. Stephen Herbert (Hastings: The Projection Box, 2004), p.29.
14. Tom Gunning, 'Never Seen This Picture Before: Muybridge in Multiplicity', in *Time Stands Still: Muybridge and the Instantaneous Photography Movement*, ed. Phillip Prodger (Stanford: Iris & B. Gerald Cantor Center for Visual Arts at Stanford University in association with Oxford University Press, 2003), p.224.
15. See: *Einstein's Miraculous Year: Five Papers That Changed the Face of Physics*, ed. John Stachel, Centenary ed. (Princeton, NJ: Princeton University Press, c1998, 2005).
16. American, 1843–1898.
17. Pepper was persuaded to do this by Thomas Eakins (1844–1916), an American realist painter who had previously used Muybridge's photographs as studies to work from. For more information on Muybridge's career as a photographer see: Braun et al., *Eadweard Muybridge: The Kingston Museum Bequest.*
18. A drawing of Londe's study was made by Paul Richer and can be found in *Phisiologie artistique de l'Homme en Mouvement.*
19. Marcel Duchamp's brother, a medical assistant, worked under Londe and exposed Duchamp to both Londe's and Marey's work. For more information see: Francois Dagognet, *Etienne-Jules Marey: A Passion for the Trace* (New York, Cambridge, MA: Zone Books; Distributed by the MIT Press, 1992). Also for an informative discussion on the influence Muybridge's studies had on subsequent art production see: David Campany, 'Moving with the times', *Tate etc*, Issue 20, Autumn 2010. Available http://www.tate.org.uk/tateetc/issue20/campanymuybridge.htm (accessed 2 February 2009)
20. José María Faerna García-Bermejo, *Marcel Duchamp*, trans. Anna Hammond (New York: Harry N. Abrahams, 1996; reprint, 2004), p.14.

21. Filmmaker Ken Jacobs stated in 1965 that filmmaking has its origins in the chronophotographs of Muybridge. See Darsie Alexander, 'Slide Show', in *Slide Show* (London: Tate, 2005). For more information on Muybridge's contribution to the history of the moving image see: Robert B. Haas, *Muybridge: Man in Motion* (Berkeley and Los Angeles: University of California Press, 1976), pp.187–203; and Gordon Hendricks, *Eadward Muybridge: The Father of the Motion Picture* (New York: Grossman Publishers, 1975).
22. Thomas, *Beauty of Another Order*, p.159.
23. Marey's photographic gun was inspired by the astronomer Pierre J. C. Janssen, who in 1874 photographed the passage of Venus between the Sun and the Earth on a light-sensitive disk, see: *Encyclopedia of Early Cinema*, p.33.
24. *A New History of Photography*, p.248.
25. John J. Videler, *Avian Flight* (New York: Oxford University Press, 2006).
26. Marey's development of the celluloid strip camera in 1889 influenced early cinema studies. Also, two research laboratories, the Station physiologique and the Institut Marey, were set up to study physiological movement. For further reading on this aspect of Marey's studies and for a fascinating resource concerning pre-cinema see: Laurent Mannoni, *The Great Art of Light and Shadow: Archaeology of the Cinema*, ed. Richard Crangle, trans. Richard Crangle (Exeter: University of Exeter Press, 2000).
27. Lisa Cartwright, *Screening the Body: Tracing Medicine's Visual Culture* (Minneapolis: University of Minnesota Press, 1995), p.36.
28. Marey at the invention of cinema told the Lumière brothers that their Cinématographe was not of interest to him because it simply re-enacted what the eye could already see and 'added nothing to the power of our sight'. For more information see David Campany, *Photography and Cinema* (London: Reaktion, 2008). Also, Dagognet, *Etienne-Jules Marey*.
29. Derrick Price, 'Surveyors and Surveyed. Photography out and About,' in *Photography: A Critical Introduction*, ed. Liz Wells (London: Routledge, 2000).
30. Jean Baudrillard, *Photography, or the Writing of Light* (CTheory.net, 2000 [cited 6 June 2006]); available from http://www.ctheory.net/articles.aspx?id=126.
31. Originally invented in 1832 by Joseph Plateau, the stroboscope used a radial disk with slits which when rotated allowed light to repeatedly pass through the disk. Edgerton invented the electronic strobe light in 1931 which used a high-speed flashing lamp to emit light. This

invention allowed Edgerton to photograph as many as 6,000 frames per second. See Harold Eugene Edgerton, *Stopping Time: The Photographs of Harold Edgerton*, ed. Estelle Jussim and Gus Kayafas (New York: H.N. Abrams, 1987).

32. Salvador Dali was enamoured with Harold Edgerton's work and its relationship to suspension and atomic matter. Dali, with the photographer Philippe Halsman, used Edgerton's work as inspiration for the famous photograph *Dali Atomicus* (1948) which shows Dali, a splash of water and cats suspended in air.
33. Marta Braun, 'The Expanded Present: Photographing Movement', in Thomas, *Beauty of Another Order*, p.150.
34. Jean Baudrillard, *The Art of Disappearance*, trans. Nicholas Zurbrugg (Brisbane: Institute of Modern Art, 1994), p.1.
35. Similar work emerged in England and was known as Vorticism. For more information on the Photofuturist movement see: *Photographie futuriste italienne*, 1911–1939, *Musee d'Art Moderne de la Ville de Paris*, 1979 and *Futurismo e Fotographie*, Milan: Multhipla, 1979. The manifesto *Futurist Photodynamism* can be found in a larger collection of Futurist manifestos, see: *Futurist Manifestos*, ed. Umbro Apollonio, trans. Umbro Apollonio (London: Thames and Hudson, 1973).
36. The Photofuturists were not accepted members of the Futurist movement. Boccioni wrote, in *Lacerba* magazine, in opposition to Bragaglia's manifesto that 'we have always refused even the most distant relation to photography with disgust and contempt, as it lies outside of art.' Boccioni in Giovanni Lista, 'The Media Heat Up: Cinema and Photography in Futurism', in *Vertigo: A Century of Multimedia Art from Futurism to the Web*, ed. Germano Celant and Gianfranco Maraniello (Milan; Bologne; New York; Skira: Museo d'Arte Moderna di Bologna; Distributed in North America by Rizzoli International Publications, 2008), p.53.
37. Anton Giulio Bragaglia 'Futurist Photodynamism 1911' in *Futurist Manifestos*, p.38.
38. *A New History of Photography*, p.254.
39. Interestingly, this was an unavoidable outcome of the early image machines, in this image it was deliberately achieved through purposeful application of photography.
40. Bragaglia, 'Futurist Photodynamism 1911', p.43.
41. Wassily Kandinsky, *Point and Line to Plane* (New York: Dover Publications, 1979). n.p.
42. Paul Virilio, *The Vision Machine* (Bloomington: Indiana University Press, 1994), p.29.

43. At times Futurist ideologies were misplaced, especially those concerning war. Importantly, however, they assigned value to the mechanical because of its apprehended ability to emancipate society from historical manifestations of power and value. The Futurists were not only aiming to depict the sensation of speed but interwoven with their belief of progress was a feeling that this progress would lead to a better future. Their utopian desires for a liberated future seem naïve today but point to a sociological attachment of desires to the mechanical, as seen in images such as *The typist.* Futurism embraced the machine and speed as part of their totalitarian vision.
44. Cinema became a possibility in 1887 when celluloid film was invented, and less than a decade later Thomas Edison had invented perforated 35mm wide film. In 1893 he patented the associated viewing device, the Kinetoscope, which was inspired by Eadweard Muybridge's Zoopraxiscope. A year later the Lumière Brothers perfected a 35mm camera and on 22 March 1895 the first public movie screening of ten short films including *Sortie des Usines Lumière à Lyon* (*Workers Leaving the Lumière Factory*) (1895) was shown. In a relatively short period of time cinema was invented and just as quickly it grew in appeal. Cinema mimicked the physiological experience of any event with the illusion of movement through time and space. Functioning in a similar way as the Kinetoscope a series of still images are presented in a successive and rapid motion. Initially this occurred without sound or colour; however, these aspects were soon developed. The earliest example of a film which combines sight and sound is by William Dickson, titled the *Dickson Experimental Sound Film*, from 1894–5. While colour was first introduced by hand painting each frame, it was later based on technology of the three colour system first discovered in 1861 by James Clerk Maxwell (1831–1879). The introduction of cinema brought about a new relationship to the light-time of photomedia and new possibilities and structures ensued. For more information concerning the early development of cinema, from the c.1890s to the c.1910s see: *Encyclopedia of Early Cinema.*
45. László Moholy-Nagy in *Light Art from Artificial Light: Light as a Medium in 20th and 21st Century Art*, ed. Peter Weibel and Gregor Jansen, trans. Peter Weibel and Gregor Jansen (Ostfildern; New York: Hatje Cantz 2006), n.p.
46. Moholy-Nagy, *Vision in Motion* (Chicago: P. Theobald, 1947).
47. László Moholy-Nagy, *New Vision, 1928, and Abstract of an Artist*, trans. Daphne M. Hoffman, 4th rev. ed. (New York: George Wittenborn, 1967).

48. Moholy-Nagy, *Vision in Motion*.
49. The English translation is *Lightplay: Black-White-Grey*.
50. The English translation is *Light-space modulator*.
51. David Cook, 'Paul Virilio: The Politics Of "Real Time"' (article A119) (Ctheory, 01/16/2003 2003 [cited 3 May 2006]); available from www.ctheory.net/text_file?360.
52. Virilio, *The Vision Machine*, p.21.
53. David Campany, 'Safety in Numbness: Some Remarks on Problems of "Late Photography",' in *Where Is the Photograph?*, ed. David Green (Maidstone, Kent; Brighton: Photoworks; Photoforum, 2003), p.129.
54. The first official postcard was printed by the Australian government in 1869. However, postcards did not reach mass circulation until the 1893 Chicago World Fair and the 1900 Paris Exposition sold postcards as souvenirs. By 1910 postcards of actors were available for movie goers so that they may have a memento of their fleeting movie experience. For more information on this see: Shelley Stamp, *Movie-Struck Girls: Women and Motion Picture Culture after the Nickelodeon* (Princeton, NJ: Princeton University Press 2000). Also, see: David Bowers, 'Souvenir Postcards and the Development of the Star System, 1912–1914', *Film History* 3, no. 1 (1989).
55. Steve Reich, 'Wavelength by Michael Snow', in *The Cinematic*, ed. David Campany (London; Cambridge, MA: Whitechapel; MIT Press, 2007), p.106.
56. This photograph of a wave is perhaps a reference to Albert Londe's chronophotograph 'wave from infront' (1903), which, as the title suggests, shows the movement of a wave from the shore line. Referenced also is the essential nature of all moving images as a series of stills, and the earliest exposition of this through chronophotography.
57. William C. Wees, *Light Moving in Time: Studies in the Visual Aesthetics of Avant-Garde Film* (Berkeley: University of California Press, 1992), p.155.
58. Paul Virilio, *Negative Horizon: An Essay in Dromoscopy*, ed. Michael Degener, trans. Michael Degener (London; New York: Continuum, 2005), p.106.
59. Chris Marker, *La Jetée*, (Nouveaux Pictures, 1962).
60. Patrick ffrench takes this analysis further, proposing that the image carries its own memory which is activiated by the viewer. See: Patrick ffrench, 'The Memory of the Image in Chris Marker's La Jetée', *French Studies*, 59, no. 1 (2005), pp.31–7.
61. John Conomos, *Mutant Media: Essays on Cinema, Video Art and New Media* (Sydney: Artspace Visual Arts Centre: Power Publications, 2007), p.190.

62. *Empire* was shot in 1964 from 8:06 pm, 25 July to 2:42 am, 26 July from 16 blocks away on the 41st floor of the Time-Life Building in New York City. The film is screened at a slower rate than it was shot so, although only 6 hours and 38 minutes of film was exposed, the film becomes 8 hours and 5 minutes long when screened.
63. Arthur and Marilouise Kroker, *Ctheory Interview with Paul Virilio: The Kosovo War Took Place in Orbital Space* (18 October 2000 [cited 21 April 2001]); available from www.ctheory.net/articles.aspx?id=132.
64. Paul Virilio, "Speed and Information: Cyberspace Alarm!," review of Reviewed Item, ibid., no. a030 (1995), http://www.ctheory.net/articles.aspx?id=72 (accessed 11 November 2008).
65. Virilio, *Negative Horizon*, p.134.
66. Kroker, *Ctheory Interview with Paul Virilio.*
67. Vilém Flusser, *Towards a Philosophy of Photography*, trans. Anthony Mathews (London: Reaktion, 2000), p.34.
68. Ibid., p.35.
69. Originally, the silver halides were held on glass using eggwhite as a binder. This provided relatively sharp images although they were easily damaged. By 1871, the problem had been solved by Dr R. L. Maddox, an amateur photographer and physician, who discovered a way to prepare gelatin dispersions of silver halides on glass plates. For more information see Dr Drew Myers, 'Chemistry of Photography.' (Place Published: The Chemical Engineers' Resource Page, 2008), http://www.cheresources.com/photochem.shtml (accessed 4 March 2009).
70. History of Kodak (cited 28 January 2009); available from http://www.kodak.com/global/en/corp/historyOfKodak/historyIntro.jhtml.
71. Naomi Rosenblum, *A World History of Photography* (New York: Abbeville Press, 1984).
72. Thomas, *Beauty of Another Order*, p.76.
73. Bernice Abbott in ibid., p.118.
74. Flusser, *Towards a Philosophy of Photography*, p.10.
75. Craig Burnett, *Jeff Wall*, ed. Tate Gallery (London: Tate Publishing, 2005), p.10.
76. Cindy Sherman, 'The Making of Untitled', in *Cindy Sherman: The Complete Untitled Film Stills* (New York: The Museum of Modern Art, 2003), p.4.
77. Sugimoto has continued this line of enquiry for decades, the entire body of work also includes outdoor drive-in movies.
78. Sugimoto in Michael Fried, *Why Photography Matters as Art as Never Before* (New Haven; London: Yale University Press, 2008).
79. Campany, *Photography and Cinema*, p.12.

80. Flusser, *Towards a Philosophy of Photography*, p.68.
81. Ibid., pp.19–20.
82. Ibid., p.20.
83. Ibid., p.20.
84. Baudrillard, *The Art of Disappearance.*
85. Flusser, *Towards a Philosophy of Photography*, p.48.
86. Magnum is a photographic cooperative founded in 1947. Its library of documentary photographs holds approximately one million images.
87. Flusser, *Towards a Philosophy of Photography*, p.48.
88. Virilio, *The Vision Machine*, p.21.
89. Ibid., p.22.
90. Henri Cartier-Bresson, *The Decisive Moment* (New York: Simon and Schuster, 1952), n.p.
91. Ibid., n.p.
92. Flusser, *Towards a Philosophy of Photography*, p.59.
93. Ibid., p.76.
94. Ibid.
95. Cartier-Bresson, *The Decisive Moment*, n.p.
96. Ibid.
97. Baudrillard, *Photography, or the Writing of Light.*
98. From the invention of paper in China, c.200 BC to nineteenth century hobbyists in Europe, collage was a craft technique. Nita Leland and Virginia Lee Williams, *Creative Collage Techniques*, 1st ed. (Cincinnati: North Light Books, 1994), p.7.
99. Sally O'Reilly, 'Collage: Diversions, Contradictions and Anomalies', in *Collage: Assembling Contemporary Art*, ed. Blanche Craig, trans. Blanche Craig (London: Black Dog Publishing, 2008), p.11.
100. Photomontage works in a similar way to collage, except that photographs or photographic negatives are cut up and re-assembled. Similar techniques can be traced back to the era of the early image machines when Henri Peach Robinson combined multiple exposures to produce composited images.
101. *A New History of Photography*, p.433.
102. John Stezaker in conversation with David Lillington in *Collage: Assembling Contemporary Art*, ed. Blanche Craig, trans. Blanche Craig (London: Black Dog Publishing, 2008), p.28.
103. Sally O'Reilly, 'Collage: Diversions, Contradictions and Anomalies', in ibid., p.8.
104. Virilio, *Negative Horizon*, p.140.
105. Alexander Alberro, 'The Dialectics of Everyday Life: Martha Rosler and the Strategy of the Decoy', in Martha Rosler, *Martha Rosler: Positions*

in the Life World, ed. M. Catherine de Zegher, trans. M. Catherine de Zegher and Gallery Ikon (Birmingham; Vienna; Cambridge, MA: Ikon Gallery; Generali Foundation; MIT Press, 1998), p.81.

106. Ian Monroe, 'Where Does One Thing End and the Next Begin', in *Collage: Assembling Contemporary Art.*
107. Michael Rush, *New Media in Art*, 2nd ed. (London: Thames & Hudson, 2005).
108. David. F Cheshire, *The Complete Book of Video*, ed. Phil Wilkinson (London: Dorling Kindersley, 1990), pp.14–5.
109. 'In camera' editing is performed while the tape is in the camera. The process relies on rewinding or advancing the tape to the desired edit point and recording. This technique requires the artist to be intimately tuned to the intricacies of their equipment and the speed of its reels to achieve the most precise cuts.
110. American, born Andrew Warhola (1928–1987).
111. Korean-born American artist (1932–2006).
112. Israeli (1932–2009).
113. Bruce Nauman's *Video Corridors* which incorporate surveillance with performance has been of significant influence to my own body of work, *Apparatus*, which will be discussed in chapter 3.
114. Paik's first video has a legendary place within video art history. As the story goes, Paik purchased a Sony Portapak in New York; having charged the battery and tested the camera he jumped in a cab to go up town. Along the way his cab ride was slowed by a papal visit which Paik attempted to shoot from his cab and that night he showed the footage to friends in a café which was claimed as the first video art piece. However, it is highly possible other artists who had access to video cameras may have shown their work sooner.
115. During the 1960s the televised image would be shown as 625 lines of light.
116. John G. Hanhardt, *The Worlds of Nam June Paik*, ed. John G Hanhardt (New York: Guggenheim Museum, 2000), p.130.
117. Rosalind Krauss, 'Video: The Aesthetics of Narcissism', *October* Spring, no. 1 (1976), p.53.
118. This approach is also seen in the photographic work of Doug and Mike Starn among others during the late part of the era.
119. *Buky Schwartz: Videoconstructions*, ed. Bill Judson (Pittsburgh: Carnegie Museum of Art; Olive Production & Publishing, 1992), p.46.
120. A reference to the construction of forms necessary for the realisation of Schwartz's work.
121. Virilio, *The Vision Machine*, p.22.

122. Anne-Sargent Wooster 'Buky Schwartz: Mirrors and Interactivity' in *Buky Schwartz: Videoconstructions*, p.18.
123. Virilio, *The Vision Machine*, p.21.
124. To make this image Klein spliced the picture of himself leaping from a window (into the arms of his Judo class) with a picture of the street and the cyclist. Klein made two version of this image, the one shown in figure 2.19 which is most often used and which Klein published in *Dimanche* in October 1960, the other is only slightly different to the first in that it shows the street sans bike rider.
125. Klein saw himself as a messiah who like Christ would transcend the Earth. He also believed he would die at the same age as Christ, about 33 or 34. Armed with this belief he tried to build a mythology for himself, through his work. Remarkably, Klein did die unexpectedly at the age of 34 from a heart attack. For more information see Sidra Stich, *Yves Klein* (Ostfildern; New York: Cantz; American distribution D.A.P. (Distributed Art Publisher), 1994.
126. Stich, *Yves Klein*, p.142.
127. Virilio, *Negative Horizon*, p.114.
128. Ibid., p.115.
129. See: section 'Slow motion: light-time in the digital era', chapter 3, where the void is discussed in terms of Flusser's photographic universe.
130. Flusser, *Towards a Philosophy of Photography*, p.51.
131. Virilio, *Negative Horizon*, p.126.
132. Essentially, Flusser's term 'magic' refers to the underlying forces of any communication; at first this magic begins in the universe of traditional images, which does not contain texts. Initially this 'magic', he claims, is full of meanings and 'gods'. Whereas in a world of traditional images which also involved hermetic texts, Flusser considers a 'prehistoric' magic to emanate subliminally.
133. Virilio, *Negative Horizon*, p.118.
134. Ibid. p.134.
135. Henri Cartier-Bresson cited in Richard D. Zakia, *Perception and Imaging*, 2nd ed. (Boston, MA: Focal Press, 2002).
136. Baudrillard, *Photography, or the Writing of Light.*

Chapter 3

1. Despite the dramatic shift of photomedia technology towards digital mechanisms, analogue technologies continue to be used, albeit in increasingly niche markets for specific aesthetic qualities. This digital era does not occur separately from analogue processes; its presence does however, alter the photomedia environment.

2. Polaroid was only a few years behind Kodak in introducing a digital camera and when they did it was aimed at the consumer market. Their offering was the PDC-200 which had a resolution of 5.6MB per image and cost US$3,000. For more information on the bankruptcy of Polaroid see: *Polaroid Instant Film Fades Out* (world news Australia, 2008 [cited 24 April 2008]); available from http://news.sbs.com.au/worldnewsaustralia/polaroid_instant_film_fades_out_540038.
3. The DSC-100 combined a digital processor mounted onto a Nikon F3 body. The DSC-100 was capable of capturing images with a resolution of 1.3 megapixels. Storage of these images was achieved via an outer module which housed batteries and a 200 megabyte hard disk drive capable of storing 156 images without compression.
4. From the early 1990s digital systems have exponentially improved storage capabilities and the resolution of images. Pixel density is mistakenly sold as 'image quality' within the consumer market, with consumers demanding more 'pixels per dollar' as a sign of superior quality and advancement of the camera. Quality in fact relies on both storage capability and lens quality for a crisp and detailed image. Barry Hendy when working at Kodak in the early 1990s coined the term 'pixel per dollar'. Interestingly he predicted that the consumer market would be content with three megapixels as that is the maximum most standard lenses in a compact camera will be able to work with; beyond this ratio data 'noise' increases.
5. The total amount of images stored relies on the amount of storage available and the size of each image captured.
6. The poem 'A dream within a dream' (1849) can be found in Edgar Allan Poe, *The Complete Tales & Poems of Edgar Allan Poe* (Edison, NJ: Castle Books, 2002).
7. The definitions and borders of such processes are blurred as one could argue that framing a shot, or cutting video or film, is part of this process.
8. See Jeff Wall, *Jeff Wall: Exposure,* ed. Jennifer Blessing, Katrin Blum and Berlin Deutsche Guggenheim (New York: Guggenheim Museum Publications, 2008), p.10.
9. Wall's method of printing his work as transparencies, which are then displayed in lightboxes is, to date, his primary presentation technique. He rarely produces works on paper but has done so for images such as *Men waiting* (2006), *War game* (2007) and *Cold storage* (2007). See: ibid.
10. Wall considered using the lightbox system for the display of his work during a long bus ride from a visit to the Prado Museum in Madrid in

which he saw the works of Spanish master painters Diego Velázquez (1599–1660) and Francisco Goya (1746–1828). At each bus terminal Wall kept seeing the luminous light boxes used for advertising and came up with an idea to combine the style of old master paintings with the modernist light box. See: Craig Burnett, *Jeff Wall*, ed. Tate Gallery (London: Tate Publishing, 2005).

11. Ralph Ellison, *Invisible Man*, Special 30th anniversary ed. (New York: Random House, 1982).
12. The others are *Odradek, Taboritska 8, Prague, July 18, 1984* (1984) which is based on Kafka's 'Dearest Father', and *After 'Spring Snow,' by Yukio Mishima, chapter 34* (2000–05), which is based on the first volume of Mishima's *Sea of Fertility* tetralogy.
13. Ralph Ellison, *Invisible Man*, pp.5–6.
14. Arthur Lubow, 'The Luminist', in *The New York Times* (New York: The New York Times, 2007), http://www.nytimes.com2007/02/25/magazine/25wall.t.html?pagewanted=all (accessed 27 April, 2009).
15. See: Jennifer Blessing, 'Jeff Wall in Black and White' in *Jeff Wall: Exposure*, pp.8–35.
16. Charles Baudelaire, *The Painter of Modern Life and Other Essays*, ed. Jonathan Mayne, trans. Jonathan Mayne, 2nd ed. (London: Phaidon, 1995).
17. Michael Fried, *Why Photography Matters as Art as Never Before* (New Haven; London: Yale University Press, 2008), p.50.
18. Russell Banks, 'Gregory Crewdson: Beneath the Roses' in Gregory Crewdson, *Beneath the Roses*, ed. Russell Banks, trans. Russell Banks (New York; London: Abrams, 2008), p.7.
19. Although Hitchcock was born in England and began his career there, he later became an American citizen and lived most of his life in California.
20. Crewdson, *Beneath the Roses*, p.7.
21. Anna Holtzman, 'Gregory Crewdson Interview with Anna Holtzman', in *The Cinematic*, ed. David Campany (London; Cambridge, MA: Whitechapel; MIT Press, 2007). Originally published as 'Interview with Gregory Crewdson', *Eyemazing* August, no. 3 (2006).
22. Fried, *Why Photography Matters as Art as Never Before*, p.50.
23. Although one could argue that Crewdson does step within the territory of the phantasmagorical and mystical with his photographic series *Twilight* (2001).
24. Poe, *The Complete Tales & Poems of Edgar Allan Poe*.
25. Gaston Bacherlard, French, 1884–1962. I am here referring to Bachelard's phenomenological study of space in which he reflects

on the relationship between physical space and, the inner and outer world of the human psyche. For more information on the elements of Bachelard's philosophy that I refer to read: Gaston Bachelard, *The Poetics of Space*, trans. Maria Jolas (Boston, MA: Beacon Press, 1969).

26. A larger paper which discusses the nature of experience in relationship to technologically mediated screen space could be written, I regrettably do not have space for it here.
27. Magic lantern shows began in at least the sixteenth century and are described in *Natural Magick*. They reached their height of popularity in the seventeenth and eighteenth centuries. For more information see: Giambattista della Porta, *Natural Magick* (London: Printed for Thomas Young and Samuel Speed, 1658).
28. Professor Erkki Huhtamo lectures at the University of California, Los Angeles.
29. Erkki Huhtamo, *Elements of Screenology: Toward an Archaeology of the Screen* (Tokyo: The Japan Society of Image Arts and Sciences, 2004).
30. Ibid., p.2.
31. This work has taken on many forms through various exhibitions. I have personally seen the work three times in a variety of galleries and art spaces. The installation I focus on in this book was seen at the exhibition *The Terminal Vision Project*, Performance Space, Eveleigh, Sydney in 2007.
32. Anna Davis, 'Imagining the Invisible', *RealTime* issue 78, April–May (2007), p.26.

    Also, this work by Beaubois bears some relation to Mike Parr's *Pushing a video camera over a hill*, 1971. In Parr's work he began with an active 16mm video camera at the base of a hill in Moore Park, New South Wales, Australia. Parr pushed the camera up the hill, the camera recording whatever was before the lens such as grass, sky and so on. For more information see: Mike Parr, *Performances 1971–2008 / Mike Parr* (Melbourne: Schwartz Media, 2008).
33. *Video Logic*, ed. Russell Storer (Sydney: Museum of Contemporary Art, 2008), p.16.
34. Milan Kundera, *Slowness* (London: Faber and Faber, 1996), p.34.
35. 'Late photography' is specifically photography which memorialises an event after it has happened. Campany uses the example of the City of New York commissioning a photographer to shoot the aftermath of the destruction of the World Trade Center in 2001. For more information see: David Campany, 'Safety in Numbness: Some Remarks on Problems of "Late Photography",' in *Where Is the Photograph?*, ed. David Green (Maidstone, Kent; Brighton: Photoworks; Photoforum, 2003).

36. Ibid., p.131.
37. Astronomer John Hershel coined the word 'snapshot' in 1860 while he was making scientific enquiries into the relationship between photography and light. See: Ann Thomas, *Beauty of Another Order: Photography in Science* (New Haven: Yale University Press in association with the National Gallery of Canada, Ottawa, 1997), p.150.
38. David Campany, 'Safety in Numbness', p.126.
39. Roland Barthes, *Camera Lucida: Reflections on Photography*, trans. Richard Howard (London: Vintage, 2000).
40. Anthony Bond, 'Bill Viola: The Tristan Project', (Sydney: Art Gallery of NSW, 2008).
41. To return to the concrete for a moment, *The fall into paradise* uses extreme high speed digital cameras, replayed in slow motion to capture footage of two divers jumping from a high springboard.
42. Robert A. F Thurman, *Essential Tibetan Buddhism* (San Francisco: Harper, 1995), p.110.
43. *The Photographic Image in Digital Culture*, ed. Martin Lister, trans. Martin Lister (London; New York: Routledge, 1995). This notion of the death of photography was widely discussed at the start of the digital era. See, for example, Timothy Druckrey, '"L'amour Faux," Digital Photography: Captured Images, Volatile Memory, New Montage', in *Exhibition Catalogue, San Francisco Camerawork* (1988), pp.4–9; Fred Ritchin, 'Photojournalism in the Age of Computers', in *The Critical Image: Essays on Contemporary Photography*, ed. Carol Squiers (San Francisco: Bay Press, 1990), pp.28–37; Anne-Marie Willis, 'Digitisation and the Living Death of Photography', in *Culture, Technology & Creativity in the Late Twentieth Century*, ed. Philip Hayward (London: John Libbey, 1990), pp.197–208; Martha Rosler, 'Image Simulations, Computer Manipulations: Some Considerations', *Ten.8* 2, no. 2 (1991). Also, William J. Mitchell, *The Reconfigured Eye: Visual Truth in the Post-Photographic Era* (Cambridge, MA: MIT Press, 1992). More recently on 22–23 April 2010 the San Francisco Museum of Modern Art held a symposium to discuss this topic. See: *Is Photography Over* (SFMOMA, 2010 [cited 1 May, 2010]); available from http://blog.sfmoma.org/tag/is-photography-over/ (accessed 1 May 2010).
44. Geoffrey Batchen, 'Phantasm: Digital Imaging in the Death of Photography', *Aperture*, no. 136 (1994), p.48.
45. Mary Ann Doane, *The Emergence of Cinematic Time: Modernity, Contingency, the Archive* (Cambridge, MA: Harvard University Press, 2002), p.10.
46. See Batchen, 'Phantasm', p.50.

47. No records of Democratis' theory exist today, also there continues to be debate as to whether this is Democratis' theory or his teacher Leucippus'; we only know of this theory through Aristotle's writing. See: Andrew Pyle, *Atomism and Its Critics: From Democritus to Newton* (Chicago: St. Augustine's Press, 1995).
48. *The Atomists, Leucippus and Democritus: Fragments: A Text and Translation with a Commentary*, ed. C.C.W. Taylor (Toronto; Buffalo: University of Toronto Press, c.1999), pp.70–1.
49. Jean Baudrillard, *The Art of Disappearance*, trans. Nicholas Zurbrugg (Brisbane: Institute of Modern Art, 1994), p.1.
50. Jean Baudrillard, *Rise of the Void Towards the Periphery* (CTheory.net, 1994 [cited 6 June 2006]); available from www.ctheory.net/text_file?pick=58.
51. Ibid.
52. Jean Baudrillard, *Photography, or the Writing of Light* (CTheory.net, 2000 [cited 6 June 2006]); available from http://www.ctheory.net/articles.aspx?id=126.
53. Paul Virilio, *The Information Bomb*, trans. Chris Turner (London: Verso, 2000).
54. Paul Virilio, *Negative Horizon: An Essay in Dromoscopy*, ed. Michael Degener, trans. Michael Degener (London; New York: Continuum, 2005), p.134.
55. Ibid., p.117.
56. Ibid., p.118.
57. Arthur Kroker. 'The Spirit of Jean Baudrillard in Memoriam: 1929–2007', in *CTHEORY: Theory, Technology and Culture*, ed. Arthur and Marilouise Kroker (CTheory, 2007), http://ctheory.net/ (accessed 8 March 2007).
58. The apparatus is defined as a fully automated apparatus which intrinsically contains so much complexity that the person playing with it cannot comprehend its mechanisms. The apparatus at discussion in this book is the photographic camera which requires human 'players', as Flusser would state. See Vilém Flusser, *Towards a Philosophy of Photography*, trans. Anthony Mathews (London: Reaktion, 2000).
59. Baudrillard, *The Art of Disappearance*, p.6.
60. Hal Foster, *The Return of the Real: The Avant-Garde at the End of the Century* (Cambridge, MA: MIT Press, 1996), p.128.
61. Melissa Miles, *The Burning Mirror: Photography in an Ambivalent Light* (North Melbourne, Vic.: Australian Scholarly Publishing, 2008), p.249.

62. Jean Baudrillard, *Symbolic Exchange and Death*, trans. Iain Hamilton Grant (London: Sage in association with *Theory, Culture & Society*, School of Health, Social and Policy Studies, University of Teesside, 1993), p.62.
63. Paul Virilio and Sylvère Lotringe, *Pure War*, trans. Mark Polizotti (New York: Semiotext(e), 1983), p.40.
64. Virilio, *Negative Horizon*.
65. French American, b.1924. Mathematician famous for his enquiries into fractal geometry.
66. Virilio, *Negative Horizon*, p.117. For the original text Virilio is referring to see: Benoit B. Mandelbrot, *Objets Fractals. Fractals: Form, Chance, and Dimension* (San Francisco: W. H. Freeman, 1977).
67. Mandelbrot coined the term 'fractal' in 1975 to refer to 'a rough or fragmented geometric shape that can be split into parts, each of which is (at least approximately) a reduced-size copy of the whole'. See Benoit B. Mandelbrot, *The Fractal Geometry of Nature* (San Francisco: W.H. Freeman, 1982).
68. Flusser, *Towards a Philosophy of Photography*, p.78.
69. Ibid., p.82.
70. Others include Lothar Hempel, Eliott Hundley, Amanda Ross-ho, Sara VanDerBeek, Alexandra Navratil and many more.
71. Tudor style was first used in medieval England (1485–1603). The style was revived in the nineteeth, twentieth and twenty-first century and in some instances is distinctively characterised in architecture by features such as external timber beams that create a stripe pattern across the architectural surface.
72. Jennifer Higgie, *David Noonan Scenes* (New York: Marc Jancou Contemporary, 2008), n.p.
73. Melissa Gronlund, 'David', *Art Review*, no. 9 (2007), p.95.
74. Johannah Fahey, 'Before and Now: The Work of David Noonan', *Eyeline* Spring, no. 58 (2005), p.44.
75. Dieter Schwarz, 'Living on the Nameless Street', in *Andro Wekua: If There Ever Was One*, ed. Kunstmuseum Winterthur (Zurich: JRP Ringier, 2006).
76. Rein Wolfs, 'Andro Wekua – a Master of Collage under the Sign of Darkness', in *If There Ever Was One*.
77. Sally O'Reilly, 'Collage: Diversions, Contradictions and Anomalies', in *Collage: Assembling Contemporary Art*, ed. Blanche Craig, trans. Blanche Craig (London: Black Dog Publishing, 2008), p.12.
78. O'Reilly, 'Collage: Diversions, Contradictions and Anomalies', p.19.

79. 'Interview with Tom Burr': http://www.youtube.com/watch?v=Sb76g-Fy8Hy4 (accessed 30 March 2008).
80. A direct reference to *2001: A Space Odyssey*. For more information on the slit-scan technique used in *2001: A Space Odyssey* see, Gene Youngblood, *Expanded Cinema*. (New York: Dutton, 1970), pp.151–6.
81. A good overview of slit-scan video works can be found at: Golan Levin, *An Informal Catalogue of Slit-Scan Video Artworks* (10 April 2005 [cited 10 April 2006]); available from http://www.flong.com/writings/lists/lists_slit_scan.html.
82. 'Time-Slice' is also registered as a technique developed by Tim MacMillan but is closer to Muybridge's technique in that a bank of cameras is set up to capture a moment. MacMillan's technique is best demonstrated in his video installation *Dead Horse* (1998). For more information and a discussion on the relationship between Time-Slice and Muybridge see: David Campany, 'Moving with the Times', *Tate etc.* issue 20, Autumn (2010), http://www.tate.org.uk/tateetc/issue20/campanymuybridge.htm (accessed 30 March 2008).
83. Kirsten Rann, 'All That Is Solid Melts', *Photofile* August–November, no. 87 (2009), p.32.
84. Laurence Simmons, 'Daniel Crooks: The Future of the Past', *Artlink* 29, no. 1 (2009).
85. Ibid., p.23.
86. Rann, 'All That Is Solid Melts', p.37.

Chapter 4

1. Kenneth Gatland and David Jefferis, *The Usborne Book of the Future: A Trip in Time to the Year 2000 and Beyond* (London: Usborne Publishing, 1979).
2. Zoë Sofia, 'Contested Zones: Futurity and Technological Art', *Leonardo* 29, no. 1 (1996).
3. Ibid., p.59.
4. Ibid.
5. 'In Another Country: Yoko Ono in Conversation with Rirkrit Tiravanija', *Artforum* Summer (2009), p.283.
6. Walter Benjamin in Maynard Solomon, *Marxism and Art: Essays Classic and Contemporary* (Detroit: Wayne State University Press, 1979), p.543.
7. Bush is most prominently known for his political role in the development of the atomic bomb for the US armed forces. Bush's essay was published in the month preceding American bombing of Hiroshima and Nagasaki which leads to the possibility that this text is a propaganda piece envisioning a better future at the hand of

potentially destructive technology. See Vanaveer Bush, 'As We May Think,' in *Multimedia: From Wagner to Virtual Reality*, ed. Ken Jordan and Randall Packer (New York: Norton, 2001). Originally published in *Atlantic Monthly* (1945) and reprinted in *Life* (1945).

8. Vannevar Bush, 'As We May Think' (11 October 2009); available from http://www.theatlantic.com.doc/194507/bush.
9. While Bush discusses capturing information in terms of linear information scanning he does not make the necessary leap to consider a cluster of complex multiple photocells capturing light.
10. See Bush, 'As We May Think'.
11. Ibid.
12. What Bush theorised was the hypertext system which is the basis of websites today. Hypertext allows instant text based linking to other texts, which is a significant advance on linear text systems.
13. For more information on Gordon Bell's *MyLifeBits* project see Gordon Bell and Jim Gemmell, 'A Digital Life', review of Reviewed Item, *Scientific American*, no. 3 (2007), http://www.scientificamerican.com/article.cfm?id=a-digital-life. Also see the website which accompanies the research project at http://research.microsoft.com/en-us/projects/mylifebits/default.aspx (accessed 22 September 2010).
14. K. Michael Hays and Dana A. Miller, *Buckminster Fuller: Starting with the Universe* (New Haven: Yale University Press, 2008).
15. *The Photographic Image in Digital Culture*, ed. Martin Lister, trans. Martin Lister (London; New York: Routledge, 1995), p.7.
16. Ibid., p.6.
17. Sofia, 'Contested Zones', p.59.
18. Ibid.
19. A prominent theory involving the instant, dramatic change in events is Catastrophe Theory, for more information see V. I. Arnold, *Catastrophe Theory*, trans. G.S. Wassermann based on a translation by R.K. Thomas, 3rd rev. and expanded ed. (Berlin; New York: Springer-Verlag, 1992).
20. Geoffrey Batchen, *Burning with Desire: The Conception of Photography* (Cambridge, MA: The MIT Press, 1997), p.216.
21. Jonathan Griffin, 'Future Conditional', *Frieze* May, no. 123 (2009).
22. Photomedia which is not specifically tied to either analogue or digital technology may potentially return to analogue in the future. One case for this can be made by researchers from the Laser Physics Centre at the Australian National University who used a modified crystal to stop light for two seconds in time. Should the technology be developed further the implications of this research include a

solid state analogue system which records light, stores it and 'replays' the light. At present, this 'replay' feature presents the stored light in reverse and can only occur once before it is destroyed. For more information see: *Stopping Light in Quantum Leap* (Physorg, 2005 [cited 23 August 2009]); available from http://www.physorg.com/news6123.html. Also see: *Freezing Light* (Science central, 2005 [cited 23 August 2009]); available from http://www.sciencentral.com/articles/view.php3?article_id=218392702.

23. Intel Corporation, founded in 1968 by Gordon E. Moore and Robert Noyce, produces microprocessors, flash memory, motherboard chipsets, network interface cards and Bluetooth chipsets.
24. For more information see *Understanding Moore's Law: Four Decades of Innovation*, ed. David C. Brock (Philadelphia: Chemical Heritage Press, 2006).
25. Computer engineer Gordon Bell in 1972 predicted a new generation of computers every decade. See C. G. Bell, R. Chen and S. Rege, 'The Effect of Technology on near Term Computer Structures', *Computer* 2, no. 5. Also more recently G. Bell, 'Bell's Law for the Birth and Death of Computer Classes', *Communications of the ACM* 51, no. 1 (2008).
26. Kurzweil is the inventor of many technological devices including voice recognition software, the Kurzweil keyboard and tools for the blind.
27. The human genome project was projected to last 15 years but due to technological advance the project was completed in 13 years. See *About the Human Genome Project* (cited 14 October 2009); available from http://www.ornl.gov/sci/techresources/Human_Genome/project/about.shtml.
28. Steve Dietz, 'Ten Dreams of Technology', *Leonardo* 35, no. 5 (2002), p.509.
29. In 1965 I. J. Good, a mathematician put forward the idea of an 'intelligence explosion' in which intelligent machines would progressively build even more intelligent machines. Later in the 1990s, computer scientist Vernor Vinge popularised the idea that human civilisation will end due to the intelligence of machines, and coined the term 'singularity' to describe this. See Vernor Vinge, *The Coming Technological Singularity: How to Survive in the Post-Human Era* (1993 [cited 28 January 2009]); available from http://www.rohan.sdsu.edu/faculty/vinge/misc/singularity.html.
30. Ray Kurzweil, *The Law of Accelerating Returns* (2001 [cited 23rd January 2007]); available from www.kurzweilai.net/articles/art0134.html, p.1.
31. I don't propose to argue Kurzweil's case here, which lies outside the scope of this book. For more information see Kurzweil's books particularly Ray Kurzweil, *The Age of Spiritual Machines: When Computers*

Exceed Human Intelligence (St Leonards: Allen & Unwin, 1999). Also, Ray Kurzweil, *The Singularity Is Near: When Humans Transcend Biology* (New York: Viking, 2006). For the most up to date information see www.kurzweilai.net.

32. Currently one capacity is privileged over the other in any device and high quality three-dimensional shooting capabilities are not available. I see this as a common feature of most consumer cameras of 2039, allowing artists the possibility of using this feature in their work. 3D photomedia has been a novelty of the medium from the era of the early image machines. Today 3D digital video is increasingly available in new release movies and is continually improved on.
33. Kate Greene, 'Photo Future', review of Reviewed Item, *MIT Technology Review,* no. 3 (2009), www.technologyreview.com/article/22460 (accessed 3 December 2009).
34. Already, a company named Lytro is developing technology which allows you to focus on aspects of an image after you have taken it. See: http://www.huffingtonpost.com/2011/06/22/lytro-camera_n_882105.html. (accessed 22 June 2011).
35. An artist has already transformed the camera into a projection device. See: Julius Von Bismarck and his Image Fulgarator, which is currently patented technology, at http://www.juliusvonbismarck.com/bank/index.php?/projects/image-fulgurator/ (accessed 3 February 2011).
36. See: Charlie Sorrel, 'The Illustrated Man: How LED Tattoos Could Make Your Skin a Screen' (2009 [cited November 20 2009]); available from http://www.wired.com/gadgetlab/2009/11/the-illustrated-man-how-led-tattoos-could-change-the-face-of-humanity (accessed 20 November 2009). Also, Katherine Bourzac, *Implantable Silicon-Silk Electronics: Biodegradable Circuits Could Enable Better Neural Interfaces and Led Tattoos* (MIT 2009 [cited 22 November 2009]); available from http://www.technologyreview.com/computing/23847/?a=f (accessed 20 November 2009).
37. Currently, technology is being developed which will allow LEDs to manipulate optical signals, creating hyper fast data transfer methods. See: http://www.guardian.co.uk/technology/2010/nov/07/bright-idea-light-bulbs-data (last accessed 18 June 2014).
38. What is now considered extremely high-resolution visual data will be the standard mode of image capture. Considering the resolutions that will be possible it will be difficult to distinguish between traditional photomedia and Computer Generated Imaging (CGI).
39. Alfred Spector, *The Intelligent Cloud* (googleblog, 2008 [cited 1 October 2008]); available from http://googleblog.blogspot.com/2008/09/intelligent-cloud.html.

40. Nathaniel Wice and Steven Daly, *Alt. Culture: An A-to-Z Guide to the '90s-Underground, Online, and over-the-Counter,* 1st ed. (Pymble: Harpercollins, 1995); Graham St. John, 'Going Feral: Authentica on the Edge of Australian Culture', *Australian Journal of Anthropology* 8, no. 2 (1997).
41. Jonathan Crary, *Techniques of the Observer: On Vision and Modernity in the Nineteenth Century* (Cambridge, MA: MIT Press, 1992).
42. Randall Packer and Ken Jordan, *Multimedia: From Wagner to Virtual Reality* (New York: Norton, 2001).
43. Jean Baudrillard, *Screened Out,* trans. Chris Turner (London: Verso, 2002).
44. Ibid. p.178.
45. Jean Baudrillard, *Symbolic Exchange and Death,* trans. Iain Hamilton Grant (London: Sage in association with *Theory, Culture & Society,* School of Health, Social and Policy Studies, University of Teesside, 1993), p.72.
46. Ibid., p.72.
47. Ibid., p.73.
48. Ibid., p.73.
49. Ibid., p.71.
50. Ibid., p.70.
51. Vilém Flusser, *Towards a Philosophy of Photography,* trans. Anthony Mathews (London: Reaktion, 2000), p.10.
52. Ibid., p.10.
53. Ibid., p.47.
54. Ibid., p.80.
55. Ibid., p.32.
56. Vilém Flusser, *Vilém Flusser: Writings,* ed. Andreas Ströhl, trans. Erik Eisel (Minneapolis: University of Minnesota Press, 2002), p.36.
57. Ibid., p.130.
58. Martha Rosler in *The Unmonumental Picture,* ed. Richard Flood et al. (London; New York: Merrell; New Museum, 2007).
59. Flusser, *Towards a Philosophy of Photography,* p.80.
60. Flusser, *Vilém Flusser: Writings,* p.40.
61. Ibid., p.41.
62. Ibid., p.41.
63. Ibid., p.131.
64. Flusser, *Towards a Philosophy of Photography,* p.93.
65. Paul Virilio's theory of global time is initiated through real-time visual technologies and the globalization of telecommunications. It therefore begins in the digital era but will be stronger in 2039. For

more information see Paul Virilio, 'Red Alert in Cyberspace', *Radical Philosophy*, no. 74, 1995.

66. Paul Virilio, *Desert Screen: War at the Speed of Light*, trans. Michael Degener (New York: Continuum, 2002), p.vii (preface).
67. Paul Virilio, *Negative Horizon: An Essay in Dromoscopy*, ed. Michael Degener, trans. Michael Degener (London; New York: Continuum, 2005), p.129.
68. *The Virilio Reader*, ed. James Der Deriam (Maldon: Blackwell Publishers, 1998), p.5.
69. For Virilio each technology has inherent in its emergence the possibility of an accident. He uses the example of the locomotive and that it has the potential to be derailed.
70. Virilio, *Negative Horizon.*
71. Ibid., p.111.
72. *The Virilio Reader*, p.72.
73. Virilio, *Negative Horizon*, p.137.
74. Ibid., p.131.
75. Peter Howe, 'Photojournalism at a Crossroads: Technology, Culture and Economics Will Determine Its Future', *Nieman Reports*, Fall 2001, p.26.
76. Tate, *Altermodern* (Tate, 2008), Video interview with Nicolas Bourriaud.
77. Carl Aigner, 'In the Eye of the Appararus', in *Missing Link: The Image of Man in Contemporary Photography*, ed. Christoph Doswald (Zurich; New York: Stemmle, 2000), p.67.
78. Florian Rötzer 'Images Within Images, or From the image to the Virtual World', *Iterations: The New Image.* ed. Timothy Druckery (New York City; Cambridge, MA: International Centre of Photography; MIT Press, 1993), p.63.
79. Thomas B. Brill, *Light: Its Interaction with Art and Antiquities* (New York; London: Plenum Press, 1980), p.vii (preface).
80. William Gibson in Julian Jonker, *Voodoo Economics: A Remix from the South, and a Requiem for Uncounted Ancestors* (2006 [cited 11 June 2006]); available from www.c.theory.net/articles.aspx?id=565.

Conclusion

1. Cubitt, Miles and Frizot provide some of the preeminent discussion in the field. See: Sean Cubitt, 'New Light', in *Transforming Aesthetics*, ed. Jill Bennett and Anna Munster (Universty Presses of New England, 2007). Also, Melissa Miles, *The Burning Mirror: Photography in an Ambivalent Light* (North Melbourne, Vic.: Australian Scholarly Publishing, 2008). And, Michel Frizot, 'Light Machines: On the Threshold of Invention', in *A New History of Photography*, ed. Michel Frizot (Cologne: Könemann, 1998).

Bibliography

'About the Human Genome Project'. In, http://www.ornl.gov/sci/techresources/Human_Genome/project/about.shtml (accessed 14 October 2009).

Aigner, Carl. 'In the Eye of the Apparatus'. In *Missing Link: The Image of Man in Contemporary Photography*, edited by Christoph Doswald. Zurich; New York: Stemmle, 2000: 63–9.

Alexander, Darsie. 'Slide Show'. In *Slide Show*, 3–32. London: Tate, 2005.

Arnold, V. I. *Catastrophe Theory*. Translated by G.S. Wassermann based on a translation by R.K. Thomas. 3rd rev. and exp. ed. Berlin; New York: Springer-Verlag, 1992.

Asimov, Isaac. *I, Robot*. London: Dennis Dobson, 1967.

The Atomists, Leucippus and Democritus: Fragments: A Text and Translation with a Commentary. Edited by C.C.W. Taylor. Toronto; Buffalo: University of Toronto Press, c1999.

Bachelard, Gaston. *The Poetics of Space*. Translated by Maria Jolas. Boston, MA: Beacon Press, 1969.

Barthes, Roland. *Camera Lucida: Reflections on Photography*. Translated by Richard Howard. London: Vintage, 2000.

Batchen, Geoffrey. 'Phantasm: Digital Imaging in the Death of Photography.' *Aperture*, no. 136 (1994): 46–51.

—. *Burning with Desire: The Conception of Photography*. Cambridge, MA: The MIT Press, 1997.

Baudelaire, Charles. *The Painter of Modern Life and Other Essays*. Translated by Jonathan Mayne. Edited by Jonathan Mayne. 2nd ed. London: Phaidon, 1995.

Baudrillard, Jean. *Simulations.* Translated by Paul Patton and Philip Beitchman Paul Foss. New York: Semiotext(e), 1983.

—. *Symbolic Exchange and Death.* Translated by Iain Hamilton Grant. London: Sage in association with Theory, Culture & Society, School of Health, Social and Policy Studies, University of Teesside, 1993.

—. *The Art of Disappearance.* Translated by Nicholas Zurbrugg. Brisbane: Institute of Modern Art, 1994.

—. 'Precession of Simulacra'. In *Simulacra and Simulation.* Ann Arbor: University of Michigan Press, 1994.

—. 'Rise of the Void Towards the Periphery' 1994. In *CTHEORY: Theory, Technology and Culture,* ed. Arthur and Mariluise Kroker. CTheory.net, www.ctheory.net/text_file?pick=58 (accessed 6 June 2006).

—. *Simulacra and Simulation.* Translated by Sheila Faria Glaser. Ann Arbor: University of Michigan Press, 1994.

—. 'Photography, or the Writing of Light' 2000. In *CTHEORY: Theory, Technology and Culture,* ed. Arthur and Marilouise Kroker. CTheory.net, http://www.ctheory.net/articles.aspx?id=126 (accessed 6 June 2006).

—. *Screened Out.* Translated by Chris Turner. London: Verso, 2002.

Bell, Gordon and Jim Gemmell. 'A Digital Life'. Review of Reviewed Item. *Scientific American,* no. 3 (2007), http://www.scientificamerican.com/article.cfm?id=a-digital-life (link no longer functional).

Bell, C. G., R. Chen and S. Rege. 'The Effect of Technology on Near Term Computer Structures'. *Computer* 5, no. 2 (1972): 29–38.

Bell, G. 'Bell's Law for the Birth and Death of Computer Classes'. *Communications of the ACM* 51, no. 1 (2008): 86–94.

Bond, Anthony. *Bill Viola: The Tristan Project.* Sydney: Art Gallery of NSW, 2008.

Bourzac, Katherine. 2009. 'Implantable Silicon-Silk Electronics: Biodegradable Circuits Could Enable Better Neural Interfaces and Led Tattoos'. In *Technology Review,* MIT, http://www.technologyreview.com/computing/23847/?a=f (accessed 22 November 2009).

Bowers, David. 'Souvenir Postcards and the Development of the Star System, 1912–1914'. *Film History* 3, no. 1 (1989): 39–45.

Braun, Marta. 'The Expanded Present: Photographing Movement'. In *Beauty of Another Order: Photography in Science.* pp.150–185. New Haven, CT: Yale University Press in association with the National Gallery of Canada, Ottawa, 1997.

Brill, Thomas B. *Light: Its Interaction with Art and Antiquities.* New York; London: Plenum Press, 1980.

Bruce, Vicki. *Visual Perception: Physiology, Psychology and Ecology.* Edited by Patrick R. Green and Mark A. Georgeson. 4th ed. Hove: Psychology, 2003.

Buerger, Janet E. *French Daguerreotypes.* Edited by International Museum of Photography at George Eastman House. Chicago: University of Chicago Press, 1989.

Buky Schwartz: Videoconstructions. Edited by Bill Judson. Pittsburgh: Carnegie Museum of Art: Olive Production & Publishing, 1992.

Burnett, Craig. *Jeff Wall.* Edited by Tate Gallery. London: Tate Publishing, 2005.

Burroughs, William. 'The Cut-up Method of Brion Gysin'. In *The New Media Reader,* edited by Nick Monfort and Noah Wardrip-Fruin pp.89–92. Cambridge, MA: MIT Press, 2003.

Bush, Vanaveer. 'As We May Think'. In *Multimedia: From Wagner to Virtual Reality,* edited by Ken Jordan and Randall Packer, 141–59. New York: Norton, 2001.

Campany, David. 'Safety in Numbness: Some Remarks on Problems of "Late Photography"'. In *Where Is the Photograph?,* edited by David Green, 123–32. Maidstone, Kent; Brighton: Photoworks; Photoforum, 2003.

—. *Photography and Cinema.* London: Reaktion, 2008.

Caporaletti, Silvana. 'Science as Nightmare: "The Machine Stops" by E.M Forster'. *Utopian Studies* 8, no. 2 (1997): 32–48.

Cartier-Bresson, Henri. *The Decisive Moment.* New York: Simon and Schuster, 1952.

Cartwright, Lisa. *Screening the Body: Tracing Medicine's Visual Culture.* Minneapolis: University of Minnesota Press, 1995.

Cheshire, David. F. *The Complete Book of Video.* Edited by Phil Wilkinson. London: Dorling Kindersley, 1990.

Clarke, Arthur C. *2001: A Space Odyssey.* Edited by Stanley Kubrick. London: Arrow Books, 1968.

Classic Essays on Photography. Translated by Alan Trachtenberg. Edited by Alan Trachtenberg. New Haven, CT: Leete's Island Books, 1980.

Clegg, Brian. *Light Years: The Extraordinary Story of Mankind's Fascination with Light.* Edited by Lizzie Hutchins. London: Piatkus, 2001.

Cloudsley, Tim. 'Romanticism and the Industrial Revolution in Britain'. *History of European Ideas* 12, no. 5 (1990): 611–35.

Collage: Assembling Contemporary Art. Translated by Blanche Craig. Edited by Blanche Craig. London: Black Dog Publishing, 2008.

Conomos, John. *Mutant Media: Essays on Cinema, Video Art and New Media.* Sydney: Artspace Visual Arts Centre: Power Publications, 2007.

Cook, David. 2003. 'Paul Virilio: The Politics of "Real Time"'. In *CTHEORY: Theory, Technolog and Culture,* edited by Arthur and Marilouise Kroker. Ctheory, www.ctheory.net/text_file?360 (accessed 03/05/06, 2006).

Crary, Jonathan. *Techniques of the Observer: On Vision and Modernity in the Nineteenth Century.* Cambridge, MA: MIT Press, 1992.

Crewdson, Gregory. *Beneath the Roses.* Translated by Russell Banks. Edited by Russell Banks. New York; London: Abrams, 2008.

Cubitt, Sean. 'New Light'. In *Transforming Aesthetics,* edited by Jill Bennett and Anna Munster, pp.168–174. University Presses of New England, 2007.

Dagognet, Francois. *Etienne-Jules Marey: A Passion for the Trace.* New York; Cambridge, MA: Zone Books; Distributed by the MIT Press, 1992.

Declaring Space: Mark Rothko, Barnett Newman, Lucio Fontana, Yves Klein. Translated by Michael Auping. Edited by Michael Auping. Fort Worth, TX; Munich; New York: Modern Art Museum of Fort Worth; Prestel, 2007.

Dietz, Steve. 'Ten Dreams of Technology'. *Leonardo* 35, no. 5 (2002): 509–22.

Doane, Mary Ann. *The Emergence of Cinematic Time: Modernity, Contingency, the Archive.* Cambridge, MA: Harvard University Press, 2002.

Domebook 2. Edited by Lloyd Kahn. Bolinas, CA: Shelter, 1971.

Druckrey, Timothy. '"L'amour Faux," Digital Photography: Captured Images, Volatile Memory, New Montage'. In Exhibition Catalogue, San Francisco Camerawork, 4–9, 1988.

Duve, Thierry de. 'Time Exposure and Snapshot: The Photograph as Paradox'. October 5, Summer (1978): 113–25.

Eder, Josef Maria. *History of Photography.* Translated by Edward Epstean. New York: Dover Publications, 1978.

Edgerton, Harold Eugene. *Stopping Time: The Photographs of Harold Edgerton.* Edited by Estelle Jussim and Gus Kayafas. New York: H.N. Abrams, 1987.

Einstein's Miraculous Year: Five Papers That Changed the Face of Physics. Edited by John Stachel. Centenary ed. Princeton, NJ: Princeton University Press, c1998, 2005.

Ellison, Ralph. *Invisible Man.* Special 30th anniversary ed. New York: Random House, 1982.

Encyclopedia of Early Cinema. Edited by Richard Abel. London; New York: Routledge, 2005.

Fahey, Johannah. 'Before and Now: The Work of David Noonan'. *Eyeline* Spring, no. 58 (2005): 42–44.

Flusser, Vilém. *The Shape of Things: A Philosophy of Design.* Translated by Anthony Mathews. London: Reaktion, 1999.

Flusser, Vilém. *Towards a Philosophy of Photography.* Translated by Anthony Mathews. London: Reaktion, 2000.

Flusser, Vilém. *Vilém Flusser: Writings.* Translated by Erik Eisel. Edited by Andreas Ströhl. Minneapolis: University of Minnesota Press, 2002.

Forster, E.M. *The Machine Stops and Other Stories.* Edited by Rod Mengham. London: André Deutsch, 1997.

Foster, Hal. *The Return of the Real: The Avant-Garde at the End of the Century.* Cambridge, MA: MIT Press, 1996.

Francblin, Catherine and Jean Baudrillard (eds). *Art and Philosophy: Baudrillard, Gadamer, Jameson, Kristeva, Lyotard, Marin, Perniola, Sloterdijk, Sollers, Virilio, West.* Translated by Catherine Francblin and Jean Baudrillard. Milan: G. Politi, 1991.

Freezing Light. 2005. In, Science central, http://www.sciencentral.com/articles/view.php3?article_id=218392702 (accessed 23 August 2009).

Fried, Michael. *Why Photography Matters as Art as Never Before.* New Haven, CT; London: Yale University Press, 2008.

Frizot, Michel. 'Light Machines: On the Threshold of Invention'. In *A New History of Photography*, edited by Michel Frizot, 15–21. Cologne: Könemann, 1998.

Frizot, Michel. 'Speed of Photography: Movement and Duration'. In *A New History of Photography*, edited by Michel Frizot, 243–57. Cologne: Könemann, 1998.

Futurist Manifestos. Translated by Umbro Apollonio. Edited by Umbro Apollonio. London: Thames and Hudson, 1973.

Galassi, Peter. *Before Photography: Painting and the Invention of Photography.* Translated by Museum of Modern Art. Edited by Museum of Modern Art. New York; Boston: Museum of Modern Art; Distributed by New York Graphic Society, 1981.

García-Bermejo, José María Faerna. *Marcel Duchamp.* Translated by Anna Hammond. New York: Harry N. Abrahams, 1996. Reprint, 2004.

Gatland, Kenneth and David Jefferis. *The Usborne Book of the Future: A Trip in Time to the Year 2000 and Beyond.* London: Usborne Publishing, 1979.

Gernsheim, Helmut. *L. J. M. Daguerre: The History of the Diorama and the Daguerreotype.* London: Secker and Warburg, 1956.

Greene, Kate. 'Photo Future'. Review of Reviewed Item. *MIT Technology Review,* no. 3 (2009), www.technologyreview.com/article/22460.

Gregory, R. L. *Eye and Brain: The Psychology of Seeing.* 4th ed. Oxford; New York: Oxford University Press, 1994.

Griffin, Jonathan. 'Future Conditional'. *Frieze* May, no. 123 (2009) pp. 170–181.

Gronlund, Melissa. 'David'. *Art Review,* no. 9 (2007): 95.

Gunning, Tom. 'Never Seen This Picture Before: Muybridge in Multiplicity'. In *Time Stands Still: Muybridge and the Instantaneous Photography Movement,* edited by Phillip Prodger, 223–28. Stanford: Iris & B. Gerald Cantor

Center for Visual Arts at Stanford University in association with Oxford University Press, 2003.

Gustavson, Todd. *Camera: A History of Photography from Daguerreotype to Digital.* New York: Sterling Publishing, 2009.

Haas, Robert B. *Muybridge: Man in Motion.* Berkeley and Los Angeles: University of California Press, 1976.

Hanhardt, John G. *The Worlds of Nam June Paik.* Edited by John G. Hanhardt. New York: Guggenheim Museum: Distributed by H.N. Abrams, 2000.

Hays, K. Michael and Dana A. Miller. *Buckminster Fuller: Starting with the Universe.* New Haven: Yale University Press, 2008.

Heinrich, Christoph. *Claude Monet, 1840–1926.* Translated by Michael Hulse. Cologne: Benedikt Taschen, 2000.

Hendricks, Gordon. *Eadward Muybridge: The Father of the Motion Picture.* New York: Grossman Publishers, 1975.

Herschel, John F. W. 'On the Chemical Action of the Rays of the Solar Spectrum on Preparations of Silver and Other Substances, Both Metallic and Non-Metallic, and on Some Photographic Processes on the Chemical Action of the Rays of the Solar Spectrum on Preparations of Silver and Other Substances, Both Metallic and Non-Metallic, and on Some Photographic Processes'. *Philosophical Transactions of the Royal Society of London* 130 (1840): 1–59.

Herschel, John F. W. 'On the Action of the Rays of the Solar Spectrum on Vegetable Colours, and on Some New Photographic Processes on the Action of the Rays of the Solar Spectrum on Vegetable Colours, and on Some New Photographic Processes'. *Philosophical Transactions of the Royal Society of London* 132 (1842): 181–214.

Hershberger, Andrew E. 'The Past, Present and Future of the History of Photography: Interviews with Peter C. Bunnell, Gretchen Garner and Britt Salvesen'. *History of Photography* 30, no. 3 (2006): 204–11.

Higgie, Jennifer. *David Noonan Scenes.* New York: Marc Jancou Contemporary, 2008.

History of Kodak. In, http://www.kodak.com/global/en/corp/historyOfKodak/historyIntro.jhtml (accessed 28 January 2009).

Hoffman, Daphne M. *Vision in Motion.* Chicago: P. Theobald, 1947.

Holtzman, Anna. 'Interview with Gregory Crewdson'. *Eyemazing* August, no. 3 (2006).

—. 'Gregory Crewdson Interview with Anna Holtzman.' In *The Cinematic,* edited by David Campany, 168–71. London; Cambridge, MA: Whitechapel; MIT Press, 2007.

Howarth, Sophie, ed. *Essays on Remarkable Photographs.* New York: Aperture, 2005.

Howe, Peter. 'Photojournalism at a Crossroads: Technology, Culture and Economics Will Determine Its Future'. Nieman Reports, Fall 2001, 25–26.

Huhtamo, Erkki. *Elements of Screenology: Toward an Archaeology of the Screen.* Tokyo: The Japan Society of Image Arts and Sciences, 2004.

'In Another Country: Yoko Ono in Conversation with Rirkrit Tiravanija.' *Artforum* Summer (2009): 281–83.

Is Photography Over. 2010. In, SFMOMA, http://blog.sfmoma.org/tag/is-photography-over/ (accessed 1 May 2010).

Jonker, Julian. 2006. 'Voodoo Economics: A Remix from the South, and a Requiem for Uncounted Ancestors'. In *CTHEORY: Theory, Technology and Culture,* edited by Arthur and Marilouise Kroker. www.c.theory.net/articles.aspx?id=565 (accessed 11 June 2006).

Jordan, Ken and Randall Packer. *Multimedia: From Wagner to Virtual Reality.* New York: Norton, 2001.

Kandinsky, Wassily. *Point and Line to Plane.* New York: Dover Publications, 1979.

Krauss, Rosalind. 'Video: The Aesthetics of Narcissism'. *October* Spring, no. 1 (1976): 50–64.

Kroker, Arthur. 'The Spirit of Jean Baudrillard in Memoriam: 1929–2007'. In *CTHEORY: Theory, Technology and Culture,* edited by Arthur and Marilouise Kroker. CTheory, 2007. http://ctheory.net/ (accessed 8 March 2007).

Kroker, Arthur and Marilouise Kroker. 2000. 'Ctheory Interview with Paul Virilio: The Kosovo War Took Place in Orbital Space'. In CTHEORY, www.ctheory.net/articles.aspx?id=132 (accessed 21 April 2001).

Kundera, Milan. *Slowness.* London: Faber and Faber, 1996.

Kurzweil, Ray. *The Age of Spiritual Machines: When Computers Exceed Human Intelligence.* St Leonards: Allen & Unwin, 1999.

Kurzweil, Ray. 'The Law of Accelerating Returns'. In, www.kurzweilai.net/articles/art0134.html, 2001 (accessed 23 January 2007).

Kurzweil, Ray. *The Singularity Is Near: When Humans Transcend Biology.* New York: Viking, 2006.

Lemagny, Jean-Claude, and André Rouillé. *A History of Photography: Social and Cultural Perspectives.* Translated by Janet Lloyd. English language ed. Cambridge; New York: Cambridge University Press, 1987.

Levin, Golan. 'An Informal Catalogue of Slit-Scan Video Artworks'. In, http://www.flong.com/writings/lists/lists_slit_scan.html, 2005 (accessed 10 April 2006).

Light Art from Artificial Light: Light as a Medium in 20th and 21st Century Art. Translated by Peter Weibel and Gregor Jansen. Edited by Peter Weibel and Gregor Jansen. Ostfildern; New York: Hatje Cantz, 2006.

Lista, Giovanni. 'The Media Heat Up: Cinema and Photography in Futurism'. In *Vertigo: A Century of Multimedia Art from Futurism to the Web*, edited by Germano Celant and Gianfranco Maraniello, 49–59. Milan; Bologne; New York: Skira; Museo d'Arte Moderna di Bologna; Distributed in North America by Rizzoli International Publications, 2008.

Litchfield, Richard Buckley. *Tom Wedgwood, the First Photographer; an Account of His Life, His Discovery and His Friendship with Samuel Taylor Coleridge, Including the Letters of Coleridge to the Wedgwoods and an Examination of Accounts of Alleged Earlier Photographic Discoveries*. London: Duckworth, 1903.

Lubow, Arthur. 'The Luminist'. In *The New York Times*, 2007. http://www.nytimes.com2007/02/25/magazine/25wall.t.html?pagewanted=all (accessed 27 April, 2009).

Mandelbrot, Benoit B. *Objects Fractals. Fractals: Form, Chance, and Dimension*. San Francisco: W. H. Freeman, 1977.

Mandelbrot, Benoit B. *The Fractal Geometry of Nature*. San Francisco: W.H. Freeman, 1982.

Manning, Peter. *Electronic and Computer Music*. 2nd ed. Oxford: Clarendon, 1993.

Mannoni, Laurent. *The Great Art of Light and Shadow: Archaeology of the Cinema*. Translated by Richard Crangle. Edited by Richard Crangle. Exeter, Devon: University of Exeter Press, 2000.

Marker, Chris. 'La Jetée'. 29 minutes: Nouveaux Pictures, 1962.

Marks, Robert W. *The Dymaxion World of Buckminster Fuller*. New York: Reinhold, 1960.

McNeil, Ian. *An Encyclopaedia of the History of Technology*. London; New York: Routledge, 1990.

Meteyard, Eliza. *A Group of Englishmen (1795 to 1815): Being Records of the Younger Wedgwoods and Their Friends, Embracing the History of the Discovery of Photography and a Facsimile of the First Photograph*. London: Longmans, Green, and Co., 1871.

Miles, Melissa. *The Burning Mirror: Photography in an Ambivalent Light*. North Melbourne, Vic.: Australian Scholarly Publishing, 2008.

Miles, Melissa. 'Focus on the Sun: The Demand for New Myths of Light in Contemporary Australian Photography'. *Australian and New Zealand Journal of Art* 2008/2009: 221–39.

Mitchell, William J. *The Reconfigured Eye: Visual Truth in the Post-Photographic Era*. Cambridge, MA: MIT Press, 1992.

Moholy-Nagy, László. *New Vision, 1928, and Abstract of an Artist*. Translated by Daphne M. Hoffman. 4th rev. ed. New York: George Wittenborn, 1967.

Monroe, Ian. 'Where Does One Thing End and the Next Begin'. In *Collage: Assembling Contemporary Art*, edited by Blanche Craig, pp.7–14. London: Black Dog Publishing, 2008.

Mulvey, Laura. *Death 24x a Second: Stillness and the Moving Image.* London: Reaktion Books, 2006.

Myers, Dr Drew. 'Chemistry of Photography'. The Chemical Engineers' Resource Page, 2008. http://www.cheresources.com/photochem.shtml (accessed 4 March 2009).

Newhall, Beaumont. *The Daguerreotype in America.* 3rd ed. New York: Dover Publications, 1976.

Newhall, Beaumont. *The History of Photography: From 1839 to the Present.* Completely rev. and enl. ed. London: Secker & Warburg, 1982.

A New History of Photography. Translated by Michel Frizot. Edited by Michel Frizot. Cologne: Könemann, 1998.

Ohlman, Herbert. 'Information: Timekeeping, Computing, Telecommunications and Audiovisual Technologies'. In *An Encyclopaedia of the History of Technology*, edited by Ian McNeil, 686–758. London; New York: Routledge, 1990.

Parr, Mike. *Performances 1971–2008.* Melbourne: Schwartz Media, 2008.

The Photographic Image in Digital Culture. Translated by Martin Lister. Edited by Martin Lister. London; New York: Routledge, 1995.

Poe, Edgar Allan. *The Complete Tales & Poems of Edgar Allan Poe.* Edison, NJ: Castle Books, 2002.

'Polaroid Instant Film Fades Out'. 2008. In, world news Australia, http://news.sbs.com.au/worldnewsaustralia/polaroid_instant_film_fades_out_540038 (accessed 24 April, 2008).

Porta, Giambattista della. *Natural Magick.* London: Printed for Thomas Young and Samuel Speed, 1658.

Price, Derrick. 'Surveyors and Surveyed. Photography Out and About'. In *Photography: A Critical Introduction*, edited by Liz Wells, 65–113. London: Routledge, 2000.

Pyle, Andrew. *Atomism and Its Critics: From Democritus to Newton.* Chicago: St. Augustine's Press, 1995.

Rann, Kirsten. 'All That Is Solid Melts'. *Photofile* August–November, no. 87 (2009): 32–37.

Reich, Steve. 'Wavelength by Michael Snow'. In *The Cinematic*, edited by David Campany, 106–07. London; Cambridge, MA: Whitechapel; MIT Press, 2007.

Rich, John. *Timber Geodesic Domes.* Auckland, NZ: Akron Consolidated, 1983.

Ritchin, Fred. 'Photojournalism in the Age of Computers'. In *The Critical Image: Essays on Contemporary Photography*, edited by Carol Squiers, 28–37. San Francisco: Bay Press, 1990.

Rosenblum, Naomi. *A World History of Photography.* 1st ed. New York: Abbeville Press, 1984.

Rosler, Martha. 'Image Simulations, Computer Manipulations: Some Considerations'. *Ten.8* 2, no. 2 (1991): 52–63.

Rosler, Martha. *Martha Rosler: Positions in the Life World.* Translated by M. Catherine de Zegher and Gallery Ikon. Edited by M. Catherine de Zegher. Birmingham, England; Vienna, Austria; Cambridge, MA: Ikon Gallery; Generali Foundation; MIT Press, 1998.

Rush, Michael. *New Media in Art.* 2nd ed. London: Thames & Hudson, 2005.

Schaaf, Larry J. *Out of the Shadows: Herschel, Talbot, and the Invention of Photography.* New Haven, CT; London: Yale University Press, 1992.

Schwarz, Dieter. 'Living on the Nameless Street'. In *Andro Wekua: If There Ever Was One,* edited by Kunstmuseum Winterthur, n.p. Zurich: JRP|Ringier, 2006.

Sherman, Cindy. 'The Making of Untitled'. In Cindy Sherman's *The Complete Untitled Film Stills.* New York: The Museum of Modern Art, 2003.

Simmons, Laurence. 'Daniel Crooks: The Future of the Past'. *Artlink* 29, no. 1 (2009): 20–27.

Sofia, Zoë. 'Contested Zones: Futurity and Technological Art'. *Leonardo* 29, no. 1 (1996): 59–66.

Solomon, Maynard. *Marxism and Art: Essays Classic and Contemporary.* Detroit: Wayne State University Press, 1979.

Sorrel, Charlie. 'The Illustrated Man: How Led Tattoos Could Make Your Skin a Screen'. In WIRED, 2009, http://www.wired.com/gadgetlab/2009/11/the-illustrated-man-how-led-tattoos-could-change-the-face-of-humanity (accessed 20 November 2009).

Spector, Alfred. 'The Intelligent Cloud'. In googleblog, 2008, ed. Karen Wickre and Alan Eagle. googleblog, http://googleblog.blogspot.com/2008/09/intelligent-cloud.html (accessed 1 October 2008).

St. John, Graham. 'Going Feral: Authentica on the Edge of Australian Culture'. *Australian Journal of Anthropology* 8, no. 2 (1997): 167–89.

Stamp, Shelley. *Movie-Struck Girls: Women and Motion Picture Culture after the Nickelodeon.* Princeton, NJ: Princeton University Press, 2000.

Stephen Herbert, Marta Braun, Paul Hill, Anne McCormack and Kingston Museum and Heritage Centre. *Eadweard Muybridge: The Kingston Museum Bequest.* Edited by Stephen Herbert. Hastings: The Projection Box, 2004.

Stich, Sidra. *Yves Klein.* Ostfildern; New York: Cantz; American distribution D.A.P. (Distributed Art Publisher), 1994.

Stopping Light in Quantum Leap. 2005. In, Physorg, http://www.physorg.com/news6123.html (accessed 23 August 2009).

Tate. *Altermodern: Tate*, 2008. Video interview with Nicolas Bourriaud.

Thomas, Ann. *Beauty of Another Order: Photography in Science*. New Haven, CT: Yale University Press in association with the National Gallery of Canada, Ottawa, 1997.

Thurman, Robert A. F. *Essential Tibetan Buddhism*. San Francisco: Harper, 1995.

Understanding Moore's Law: Four Decades of Innovation. Edited by David C. Brock. Philadelphia: Chemical Heritage Press, 2006.

The Unmonumental Picture. Edited by Richard Flood, Massimiliano Gioni, Laura J. Hoptman, Lisa Phillips and New Museum. London; New York: Merrell; New Museum, 2007.

Van Nostrand's Scientific Encyclopedia. Edited by Douglas M. Considine. 7th ed. New York: Van Nostrand Reinhold, 1989.

Videler, John J. *Avian Flight*. New York: Oxford University Press, 2006.

Video Logic. Edited by Russell Storer. Sydney: Museum of Contemporary Art, 2008.

Vinge, Vernor. 1993. 'The Coming Technological Singularity: How to Survive in the Post-Human Era'. In, http://www.rohan.sdsu.edu/faculty/vinge/misc/singularity.html (accessed 28 January, 2009).

Virilio, Paul. *The Aesthetics of Disappearance*. Translated by Philip Beitchman. 1st English ed. New York: Semiotext(e) Books, 1991.

—. *The Vision Machine*. Bloomington: Indiana University Press, 1994.

—. 'Red Alert in Cyberspace'. *Radical Philosophy*, no. 74 (1995): 2–4.

—. 'Speed and Information: Cyberspace Alarm!' Review of Reviewed Item. Ctheory, no. a030 (1995), http://www.ctheory.net/articles.aspx?id=72.

—. *The Virilio Reader*. Malden, MA: Blackwell Publishers, 1998.

—. *The Information Bomb*. Translated by Chris Turner. London: Verso, 2000.

—. *Desert Screen: War at the Speed of Light*. Translated by Michael Degener. New York: Continuum, 2002.

—. *Negative Horizon: An Essay in Dromoscopy*. Translated by Michael Degener. Edited by Michael Degener. London; New York: Continuum, 2005.

—. *Art as Far as the Eye Can See*. Translated by Julie Rose. English ed. New York: Berg, 2007.

Virilio, Paul and Sylvere Lotringe. *Pure War*. Translated by Mark Polizotti. New York: Semiotext(e), 1983.

The Virilio Reader. Edited by James Der Deriam. Maldon, MA: Blackwell Publishers, 1998.

Wall, Jeff. *Jeff Wall: Exposure.* Edited by Jennifer Blessing, Katrin Blum and Berlin Deutsche Guggenheim. New York: Guggenheim Museum Publications, 2008.

Ware, Mike. 'On Proto-Photography and the Shroud of Turin'. *History of Photography* 21, no. 4 (1997): 261–69.

Warner Marien, Mary. *Photography: A Cultural History.* London: Laurence King, 2002.

Wedgwood, B. and H. Wedgwood. *The Wedgwood Circle, 1730–1897, Four Generations of a Family and Their Friends.* Westfield NJ: Eastview Editions, 1980.

Wees, William C. *Light Moving in Time: Studies in the Visual Aesthetics of Avant-Garde Film.* Berkeley: University of California Press, 1992.

Wice, Nathaniel and Steven Daly. *Alt. Culture: An A-to-Z Guide to the '90s-Underground, Online, and over-the-Counter.* 1st ed. Pymble: Harpercollins, 1995.

Wilder, Kelley E. 'Disdéri, André-Adolphe-Eugène'. In *The Oxford Companion to the Photograph*, edited by Robin Lenman pp.742. Oxford: Oxford University Press, 2005.

Williams, Nita Leland and Virginia Lee. *Creative Collage Techniques.* 1st ed. Cincinnati: North Light Books, 1994.

Willis, Anne-Marie. 'Digitisation and the Living Death of Photography'. In *Culture, Technology & Creativity in the Late Twentieth Century*, edited by Philip Hayward, 197–208. London: John Libbey, 1990.

Wolfs, Rein. 'Andro Wekua – a Master of Collage under the Sign of Darkness'. In *Andro Wekua – If There Ever Was One.* pp.9–124. Zurich: JPR Ringier, 2006.

Youngblood, Gene. *Expanded Cinema.* 1st ed. New York: Dutton, 1970.

Zakia, Richard D. *Perception and Imaging.* 2nd ed. Boston, MA: Focal Press, 2002.

INDEX

For Product Safety Concerns and Information please contact our EU representative GPSR@taylorandfrancis.com
Taylor & Francis Verlag GmbH, Kaufingerstraße 24, 80331 München, Germany

www.ingramcontent.com/pod-product-compliance
Lightning Source LLC
LaVergne TN
LVHW050640100826
845148LV00011B/1921